Mines, Minerals & Metallurgy

by

D. D. Hogarth
University of Ottawa

P. W. Boreham
Dartford Borough Museum

J. G. Mitchell
University of Newcastle upon Tyne

Mercury Series
Directorate
Paper 7

published by the
Canadian Museum of Civilization

with the authorization of the
Meta Incognita Project Steering Committee

CANADIAN CATALOGUING IN PUBLICATION DATA

Hogarth, D.D.

Martin Frobisher's northwest venture, 1576-1581: mines, minerals & metallurgy

(Mercury series)
(Directorate paper / Canadian Museum of Civilization; no. 7)
Includes an abstract in French.
Includes bibliographical references.
ISBN 0-660-14018-7

1. Gold ores – Northwest Territories – Baffin Island – History. 2. Ore deposits – Northwest Territories – Baffin Island – History. 3. Metallurgy – Northwest Territories – Baffin Island – History. 4. Mineralogy – Northwest Territories – Baffin Island – History. 5. Baffin Island (N.W.T.) – Discovery and exploration. I. Boreham, P.W. (Peter W.), 1955- II. Mitchell, J.G. (John G.), 1943– III. Canadian Museum of Civilization. IV. Title. V. Series. VI. Series: Directorate paper (Canadian Museum of Civilization); no. 7.

QE390.2.G65H63 1994   549.9719'5   C93-099716-6

PRINTED IN CANADA

Published by
Canadian Museum of Civilization
100 Laurier Street
P.O. Box 3100, Station B
Hull, Quebec
J8X 4H2

EDITORIAL COORDINATOR: Stephen Alsford

HEAD OF PRODUCTION: Deborah Brownrigg

PRODUCTION OFFICER: Lise Rochefort

COVER DESIGN: Purich Design Studio

COVER PHOTO: Martin Frobisher, portrayed by Dutch artist Cornelis Ketel. Painted before the departure of the second voyage, this portrait cost the Cathay Company £5 and is now in the Bodleian Library, Oxford, as Poole Portrait 50. Published with permission.

OBJECT OF THE MERCURY SERIES

The Mercury Series is designed to permit the rapid dissemination of information pertaining to the disciplines in which the Canadian Museum of Civilization is active. Considered an important reference by the scientific community, the Mercury Series comprises over three hundred specialized publications on Canada's history and prehistory.

Because of its specialized audience, the series consists largely of monographs published in the language of the author.

In the interest of making information available quickly, normal production procedures have been abbreviated. As a result, grammatical and typographical errors may occur. Your indulgence is requested.

Titles in the Mercury Series can be obtained by writing to:

Mail Order Services
Publishing Division
Canadian Museum of Civilization
100 Laurier Street
P.O. Box 3100, Station B
Hull, Quebec
J8X 4H2

BUT DE LA COLLECTION

La collection Mercure vise à diffuser rapidement le résultat de travaux dans les disciplines qui relèvent des sphères d'activités du Musée canadien des civilisations. Considérée comme un apport important dans la communauté scientifique, la collection Mercure présente plus de trois cents publications spécialisées portant sur l'héritage canadien préhistorique et historique.

Comme la collection s'adresse à un public spécialisé, celle-ci est constituée essentiellement de monographies publiées dans la langue des auteurs.

Pour assurer la prompte distribution des exemplaires imprimés, les étapes de l'édition ont été abrégées. En conséquence, certaines coquilles ou fautes de grammaire peuvent subsister : c'est pourquoi nous réclamons votre indulgence.

Vous pouvez vous procurer la liste des titres parus dans la collection Mercure en écrivant au :

Service des commandes postales
Division de l'édition
Musée canadien des civilisations
100, rue Laurier
C.P. 3100, succursale B
Hull (Québec)
J8X 4H2

Canada

Abstract

The aim of Martin Frobisher's first voyage (1576) was to find a Northwest Passage to Cathay, but after reports of finding gold, interest shifted to mining. During the second voyage (1577), after which Britain laid claim to "Meta Incognita" (Baffin Island), a mine was opened near southeast Baffin Island, and 158 tons of ore were shipped to Bristol and London. This first mine, and six others nearby, were worked in the last voyage (1578), which resulted in 1136 tons of ore being sent to Dartford. One ship carrying another 110 tons was wrecked and beached in Smerwick Harbour, Ireland.

Furnaces in London produced high-grade assays, but they were abandoned in 1579. Later that same year the newly constructed Dartford works extracted a little silver, but the enterprise soon collapsed. Although Michael Lok, its creator, went to jail, Frobisher and his captains became famous.

Frobisher's ore consisted of metamorphosed mafic and ultramafic rocks characterized by hornblende, unusual textures, an uncommon chemical composition (high iron and aluminum, significant chromium and nickel), and two ages (1470 and 1840 million years). The percentage of gold was phenomenally low, commonly approaching the abundance in the earth's crust. The spectacular grade reported in 1577-78 may have been due to incompetent assayers or gold and silver added deliberately to the furnace charge.

Résumé

Martin Frobisher entreprit son premier voyage (1576) dans l'intention de découvrir un passage, par l'ouest, qui mènerait à la Chine, mais après avoir trouvé de l'or, l'intérêt se porta sur l'extraction du minerai. Lors de son deuxième voyage (1577), après lequel la Grande-Bretagne a revendiqué la possession de Meta Incognita (l'île de Baffin), une mine fut aménagée au sud-est de l'île et on a envoyé 158 tonnes de minerai à Bristol et à Londres. Cette première mine, de même que six autres – ouvertes non loin de là –, furent exploitées lors de sa dernière expédition (1578), où 1136 tonnes de minerai prirent le chemin de Dartford. Un bateau qui transportait 110 tonnes supplémentaires fit naufrage et échoua dans la baie de Smerwick, en Irlande.

Des fonderies de Londres ont produit du métal à haute teneur en or, mais elles ont été fermées en 1579. Plus tard la même année, la toute nouvelle fonderie Dartford parvint à extraire un peu d'argent, mais l'entreprise ne tarda pas à sombrer. Alors que Michael Lok, fondateur de l'établissement, était emprisonné, Frobisher et ses capitaines passèrent à l'histoire.

Le minerai de Frobisher était composé de roches métamorphiques, mafiques et ultramafiques, qui étaient caractérisées par la présence de la hornblende, par des textures anormales, par une composition chimique atypique (forte concentration de fer et d'aluminium, concentration prononcée de chrome et de nickel), et qui comprenaient des unités de deux âges distincts (1470 et 1840 millions d'années). La teneur en or était remarquablement faible, voisine de ce qui se trouve en général dans la croûte terrestre. La teneur mirobolante rapportée de 1577 à 1578 était peut-être due à l'incompétence des essayeurs ou à un ajout intentionnel d'or et d'argent à la charge placée dans le fourneau.

CONTENTS

LIST OF TABLES

LIST OF FIGURES

Preface

This is an appraisal of Martin Frobisher's voyages to Baffin Island in 1576, 1577 and 1578, from the viewpoint of his attempt to mine and process gold ore. We have examined the literature and have selected various items for close scrutiny and, wherever possible, have consulted the original manuscripts. Previously, the voyages have been looked at from the perspective of arctic exploration and anthropology. Here mining and metallurgical history are our concern.

In 1577 and 1578, Frobisher brought to England a total of more than 1000 tons of rock that went through many 'proofs' for precious metals. The initial assays showed considerable gold, succeeding ones somewhat less, and the final tests showed virtually no gold at all. What is generally not known is that Frobisher put some of his rock to pilot plant production where from most runs, gold and silver were actually recovered, albeit a small amount but considerably more than would be predicted from the latest assays (and the earlier ones from 'reputable' goldsmiths). How do we explain this inconsistency?

This leads us to other considerations. For example, what exactly was that so-called 'black ore'? Was it distinctive mineralogically, texturally and in age from rocks nearby and elsewhere? Did the various mines supply more than one type of ore? How much gold and silver was really in it? These questions have concerned writers in the past, but they were handicapped by lack of samples. How does one recover specimens that were tested in the furnace over 400 years ago, especially where the furnace site, after repeated disturbances, cannot be identified? Of course, the remains are lost forever. But fortunately three sites are available, from which we have obtained specimens: (1) the storage heaps at three mines in the Baffin Island area, worked in 1577 and 1578 but today still holding some loose pieces; (2) the central part of the Borough of Dartford, Kent, where some 1000-odd tons were eventually piled and then discarded; and (3) a beach in Ireland, near the wreckage of a Frobisher vessel that failed to reach home port.

A further problem emerges. After the final voyage, which took place in 1578, some of the ore mysteriously disappeared. We trace it across the Atlantic in the *Emanuel*, learn of its removal from the wrecked vessel in Smerwick Harbour, Ireland, and picture it brought to safety on the shore. Then, in 1579, it is gone. What happened? Some 'black ore' can still be found on the beach, but only after considerable effort. Most has vanished. Two possible explanations have hitherto been proposed: either the ore disappeared through the natural forces of erosion or else it was retrieved by the ship owner, Richard Newton. Now, after a new and minute investigation of the beach at Smerwick Harbour, coupled with the study of additional cobbles and pebbles, we formulate a new theory.

All of these facts will be collated in an attempt to present an integrated story of what really happened between 1576 and 1581: the voyages, the mining, the testing, and the metal production. Operations in this six-year period will be referred to as 'the Frobisher northwest venture' or 'the Frobisher enterprise'. Then we will consider the effects of this enterprise, both from the short- and long-range point of view.

One hundred and twenty three specimens of Frobisher's 'black ore' have been examined in this study, from Baffin Island and vicinity (Canada), Dartford (England) and Smerwick Harbour (Ireland).

Acknowledgements

The investigation of 'black ores' from Dartford was facilitated through the loan of specimens by Mr C. Baker (Dartford District Archaeological Group) and Dr A.M. Clark (Natural History Museum). Two thin sections of Dartford 'black ore' in the Harker Collection were examined through the kindness of Dr G.A. Chinner (Cambridge University). Four specimens of 'black ore' from Baffin Island area were provided from the Kenyon collection at the Department of New World Archaeology, Royal Ontario Museum.

We would like to thank Messers R. Hartree and J. Loop (University of Ottawa) for X-ray spectrographic, atomic absorption and direct-current plasma analyses of samples, Dr C.T. Williams (Natural History Museum) for instrumental neutron activation analyses, Mr. P. Jones (Carleton University) for electron microprobe analyses, Mr E.W. Hearn (University of Ottawa) for line drawings and photography, Ms R.H. Moore and Mr D.G. Harris for Latin and Danish translations, respectively, Ms G.C. Jones and Mr K. Telmer for computer language translation, and Ms S. Downing, J. Hayes and M. Low for typing. Parts of the manuscript were read by Drs K. Bell, R.W. Boyle, A.D. Fowler, R. Hutchison, A.E. Lalonde, R. Kretz, R. McGhee, D.J. McLaren, C.J. Stanley and R.F. Symes. Information on Elizabethan mining was supplied by Mr L. Willies (Peak District Mining Museum, U.K.). The authors wish to thank the editor of the Mercury Series, Mr S. Alsford, for many helpful suggestions. We are specially indebted to Mr D.T. Moore (formely NHM) for assistance in preparing an early draft of the manuscript and providing neutron activation and electron microprobe analyses.

We also would like to thank the staff of various libraries and archival collections for their valuable assistance in locating books, periodicals and manuscripts, and interpreting Elizabethan secretarial hand, especially the personnel of the Archiepiscopal Library of Lambeth, Archivo Segreto Vaticano, British Library, Cornwall Record Office, Guildhall Library, Hatfield House, Huntingdon Library, McMaster University Library, National Archives of Canada, New York Public Library, Pepys Library, Public Record Office, Royal Irish Academy, Somerset Record Office, and Weymouth Borough Archives. Permission to publish illustrations was granted by the Bodleian Library (Frontispiece), British Library (Figs. 3, 4a), British Museum (Fig. 7), Department of Energy, Mines and Resources of Canada (Fig. 8), École Nationale Supérieure des Beaux-arts, Paris (Fig. 4b), National Archives of Canada (Fig. 36), National Maritime Museum (Fig. 11), Public Record Office (Figs. 2, 6, 15, 16, 19), and Rochester Bridge Trust (Fig. 18).

One of us (D.D.H.) would like to acknowledge The Natural History Museum for facilities and assistance during sabbatical leave, 1988-9. He is particularly indebted to the former Keeper of Mineralogy, Dr A.C. Bishop, without whose encouragement this research would not have taken place. Transportation in Frobisher Bay, Baffin Island, was via the longliner *Pitsiulak* and D.D.H. is grateful to Dr W.W. Fitzhugh (Smithsonian Institution) for accommodation with his field parties in 1990, 1991 and 1992. In 1991 he was ably assisted on Baffin Island by Davin Ala and in 1992 by André Gonciar (undergraduate students, University of Ottawa). This publication was authorized and funding provided by the Meta Incognita Project Steering Committee.

Abbreviations Used

ALL	Archiepiscopal Library of Lambeth (London)
APC	Acts of the Privy Council
ASV	Archivo Segreto Vaticano (Rome)
BL	British Library, Department of Mineralogy (London)
BM	The Natural History Museum, Department of Mineralogy (London)
BRO	Bristol Record Office (Bristol)
BUAM	Michaud, J.F. [Editor] 1811-1828. *Biographie universelle, ancienne et moderne.* Paris (Chez Michaud Frères)
CP	Cecil Papers (Hatfield House)
CRO	Cornwall Record Office (Truro)
DAB	Johnson, A. & Malone, D. [Editors] 1927-1936. *Dictionary of American Biography.* New-York (Charles Scribner's Sons)
DBI	Ghisalberti, A.M. [Editor-in-chief] 1960-. *Dizionario biografico delgi Italiani.* Rome (Instituto della Enciclopedia Italiana)
DHF	Decaux, A. & Castelot, A. [Editors] 1981. *Dictionnaire d'histoire de France Perrin.* Paris (Librarie Académique Perrin)
DNB	Stephen, L. & Lee, S. [Editors] 1885-1900. *Dictionary of National Biography.* London (Smith, Elder & Co.)
HL	Huntington Library (San Marino)
NAC	National Archives of Canada (Ottawa)
NBG	Hoefer, J.F.C. [Editor] 1852-1866. *Nouvelle biographie générale.* Paris (Firmin Didot Frères, Fils et Cie)

NDB	Historische Kommission der Bayerischen Akademie der Wissenschaften [Editors] 1953 -. *Neue Deutsche Biographie*. Berlin (Duncker & Humblot)
NLW	National Library of Wales (Aberystwyth)
NMM	National Maritime Museum (Greenwich)
OCSL	Ward, P. [Editor] 1978. *Oxford Companion to Spanish Literature*. Oxford (Clarendon Press)
OED	*The Oxford English Dictionary*. Second edition. 1989. Oxford (Oxford)
PL	Pepys Library (Magdalene College, Cambridge)
PRO	Public Record Office (London)
RIA	Royal Irish Academy (Dublin)
SM	Stefansson, V. & McCaskill, E. [Compilers & Editors] *The three voyages of Martin Frobisher*. London (The Argonaut Press)
SRO	Somerset Record Office (Taunton)

Units and values of 1578

Tons were probable long tons, approximately equal to metric tonnes. Gold was worth £3 per ounce (troy), silver 5s. 2d per ounce (troy).

To convert	to	multiply by
Chaldrons (or 'chalders') coal (Newcastle scale)	tons (long)	1.5
tons ('T', long)	hundredweights	20
hundredweights ('cwt')	quarters	4
quarters ('qz')	pounds (avdp)	28
pounds ('lb', avdp)	hundredweights	0.00893
pounds (avdp)	kilograms	0.454
ounces ('oz' troy)	pounds (avdp)	0.0686
ounces ('oz', troy) per ton (long)	ppm	30.6
loths per hundredweight	ounces (troy) per ton (long)	10
pennyweights ('dwt', troy)	ounces (troy)	0.05
grains ('gr', troy)	ounces (troy)	0.002083
leagues (naut.)	miles (statute)	3.455
miles	kilometres	1.609
feet	metres	0.305
werckschuh	metres	0.285
palms	centimetres	7.6
digits	centimetres	1.905

Throughout the text, years have been converted to the post-1753 (Gregorian) calendar.

CHAPTER 1
Introduction

Man has gone to the ends of the earth in search of gold and suffered many hardships in his quest. At Colchis he battled the fire-breathing dragon, on the Orinoco he pressed through mosquito-infested swamps, in Peru he crossed the snow-capped Andes, across South Australia he trekked the sun-parched plains, and to reach the Klondyke he climbed the icy slopes of Chilkoot Pass. Martin Frobisher's second and third voyages belong in this class: just 85 years after Columbus, his tiny ships anchored in the tidal eddies off Baffin Island in arctic Canada, where rich black gold ore was thought to abound. Now after a lapse of four centuries, it is interesting to re-examine the details and consider the overall effects of Frobisher's extraordinary effort. In this chapter we will consider further the objectives of the Frobisher northwest venture. We will also regard the rapid circulation of news of the voyages and note certain items of information that were concealed from the public and have remained as unpublished manuscripts to this day. Then we will briefly describe mining in the British Isles at the time when Frobisher made his Baffin debut. And finally we will draw attention to certain persons involved, to various degrees, in the enterprise itself - all background for our main theme: Frobisher's mines, minerals and metallurgy, 1576-1581.

Frobisher was born neither miner nor maritimer. His father was Barnard Frobisher, heir to the family estate at Altofts, a rural community in western Yorkshire. Barnard was the local squire, born about 1505, appointed churchwarden at Normanton and bailiff of Hetton (Kirkheaton) and Battley. His mother was Margaret daughter of Sir John York, then Master of the Mint at the Tower of London. Young Martin was orphaned in 1549 and at the early age of 14, was adopted by his uncle Sir John York of Fountains Abbey, Gowthwaite, Kent. Here Martin yearned for a maritime life and in 1553, through Sir John's connections, he was placed on the Wyndham expedition to Guinea.[1] Martin took to seamanship wholeheartedly and became an indomitable sailor with a special flair for adventure. Still in his twenties, he nourished an ambition to discover and explore the Northwest Passage, a new and shorter route to the Orient where immense riches were believed to lie. It was, as he claimed, "the onely thing in thing of the Worlde that was left yet undone" (George Best in *SM* 1: 46). In his opinion, he had been born too late.

However, at this time, the search for a new route to Cathay, either via the east or west, was, by Royal Charter, vested in the Muscovy Company. It was only through persistence of a powerful group of dissidents in the company, led by their treasurer Michael Lok, that the privilege, in 1575, was finally relinquished. Lok and Frobisher then set out to arrange financing for an expedition to the northwest, Lok putting up £100 from his own pocket and collecting £145 from fellow Londoners, Frobisher raising £300 at the Court of Elizabeth (Manhart 1924). The total fell short of the target but was sufficient to field a modest voyage in 1576.

Who were these sponsors or 'venturers'? Of the 18 charter members, at least 10 were shareholders of the Muscovy Company (McDermott 1984: 36, 42-43), including the renouned Lionel Ducket (a governor), Thomas Randolph (a former ambassador to the Russian court), Anthony Jenkinson (England's most famous explorer and trader in Russia and the Middle East) and, of course, Michael Lok (their London agent and treasurer). Regarded from another angle the group was split almost equally between Court and City. Their aim was simply to make a healthy profit on their investment. But of these early 'venturers', it was Lok who provided the driving energy for the project: Lok the organizer and Lok the ever-active participant. Frobisher was his willing accomplice.

Gold mines were far from Frobisher's mind when he embarked for the northwest in 1576. At that time the Arctic was regarded as an unfavourable locale for the occurrence of precious metals.[2] True he had on board a touchstone,[3] but this was probably to test the purity of gold in the Orient. The ultimate aim in this voyage was to find the Northwest Passage, but from the landfall the first specimen of 'black ore' was collected, later reported to contain large amounts of gold and silver. The news of precious metals came as a complete surprise.

Early in 1577, after this news leaked out, emphasis quickly changed to minerals but the immediate response of investors was negative: five 'venturers' from the first voyage and twenty potential investors in the second withdrew their support (McDermott 1984: 46). New shareholders were not obviously oriented to gold and mining: only Sir William Winter, Sir Julius Caesar and the Earl of Pembroke were shareholders in other mining ventures but they were not at all involved in the administration of these companies. In fact, English mine developers generally ignored the project. However, for royalty and nobility investment increased sustantially, with backing led by the Queen, who donated the galliass *Ayde* (successively valued at 500, 750 and £850) and added £250 to the parcel, which, after assessments, amounted to a grand total of £4000. Later (1578) the Earl of Oxford, notoriously impetuous and irresponsible in money-matters (Sidney Lee in *DNB* **58**: 227), acquired half of Michael Lok's immense holdings and was finally assessed at £2520.[4] By and large, investment came from the wealthy, who could well afford to speculate in a high risk venture.

On return of the second voyage (fall 1577) the Queen named the islands within and land surrounding Frobisher Bay ('Frobisher's Strait') *Meta Incognita*, Latin for unknown limits, "as a marke and bounds utterly hitherto unknown" (G. Best in *SM* **1**: 80). Then, at the request of Queen Elizabeth, Dr. Dee officially inscribed the claim of 'Title Royall' on parchment, for lands that extended from Florida northward along the Atlantic seaboard, past Newfoundland, along the Labrador coast, across Hudson Strait to Baffin Island, eastward to Greenland, Iceland, Kolguyev, Novaya Zemlya and Voygach. This gigantic land claim was based on priority of European discovery according to eleven voyages (six of which were legendary).[5] Needless to say this land claim, in its entirety, was never pursued, but England clung doggedly to Baffin, Labrador and Newfoundland, with bonds strengthened by later voyages.

The Frobisher literature

The northwest voyages of Martin Frobisher were, up to about 1700, the best documented of all voyages of arctic exploration. Three books left the press shortly after return of the ships: Settle (1577) describing the second voyage, Ellis (1578) the third, and Best (1578) all three voyages. Settle's account, with realistic descriptions of a far-away, inaccessible and exotic country and its distinctive inhabitants, immediately caught the attention of the public. It became a 'best seller' and went through editions in French and Latin, as well as English[6] (*SM 2*: 224-7). There followed a German edition (Settle 1580), a nineteenth century English edition (Settle 1868), and a twentieth century English facsimile (Settle 1969). The woodcut (Fig. 1), included with the French, Latin and German editions, suggests a sketch made by someone with artistic talent, but with limited knowledge of the country or its people.[7]

The Ellis volume (*SM 2*: 27-51) survives as a single holding (in the Huntington Library), although it has been widely quoted in the past and has been reprinted in facsimile (Ellis 1922). It is best remembered for its woodcut showing four views of the same large, decaying iceberg. Similar sights have impressed every summer visitor to the Frobisher Bay area. Best's volume (*SM 1*: 1-129) is the most widely known, and has been regarded as the official narrative of the voyages.[8] It is, however, the least reliable account of the first and third voyages (for which we have ships' logs). It is even doubtful that Best was present on the first voyage. His name does not appear in the relevant Exchequer book (*PRO*, E164/35) and his account seems very much 'second hand'. A curious woodcut of a 'sea unicorn' [narwhal] and two roughly drawn maps are included in this volume.

Another contemporary book deserves mention: *Scoprimento dello Stretto Artico et di Meta Incognita* [the discovery of the Arctic Strait and Meta Incognita] (Anania 1582). It has not been described previously. The book is divided into two parts: the first is the account of the second voyage somewhat modified after Settle, and the second an account of the third voyage, modified and augmented after Ellis. Illustrations are woodcuts pressed from new blocks copied from Settle's Inuit and Ellis' icebergs.[9]

All literature on the voyages, available at the end of the sixteenth century was printed in Hakluyt (1598-1600).[10] This comprises the books by Settle, Ellis, and Best, and various manuscripts including the journal of Christopher Hall (first voyage) and notes of Thomas Wiars (third voyage). These were republished and combined with the most relevant State Papers by Collinson (1867). Stefansson & McCaskill (1938) augmented

Figure 1: Illustrations from the French (upper) and Latin (lower) editions of Settle. The continental landscape, presence of trees, Franco-German styled tents and somewhat European attire suggest the original artist was not present on the voyages. Kayaks and the use of dogs were familiar from descriptions and materials returned by Frobisher.

Table 1: *Ten manuscripts dealing with the voyages, assays and furnaces*

No	Period	Abbreviated title	Author(s)	Loc.	Identification	Leaves
1	1575-8	Accounts of Michael Lok (and others)	M. Lok *et al.*	PRO	E 164/35	152*
2	1575-81	Testimony of Michael Lok	M. Lok	BL	Cotton, Otho E 8/8	13‡
3	1576-9	The doynges of Captayne furbisher	[M. Lok]	BL	Lansdowne 100/1	15
4	1577-8	Tryeing of ye Northwest Ewre	[M. Lok]	PRO	SP 12/122/62	6
5	1578	Thaccount of Michael Lok	M. Lok	HL	HM 715	53**
6	1578-80	Accounts of Michael Lok (and others)	M. Lok *et al.*	PRO	E 164/36	176*
7	1578	The Journal of Captain Edward Fenton	E. Fenton	PL	2133	35+
8	1578	Notes on the Voyage to Meta incognita	E. Sellman	BL	Harley 167/40	16
9	1578	The voyage ... in ye Judith	Anon.	BL	Harley 167/41	2‡
10	1578	The accownt of the third Voyage	C. Hall	BL	Harley 167/42	18

* 36 leaves blank, ** 26 leaves blank, + 2 leaves blank, ‡ damaged by fire.

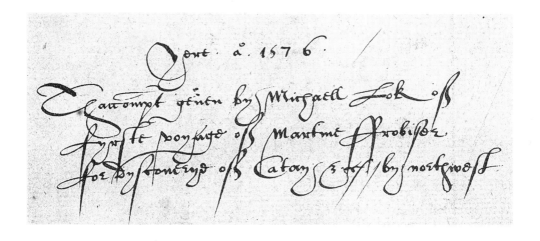

Figure 2: Title page of expense account for the first voyage (MS1, Table 1).
Transcription:

Yere. aº. 1576.
Thaccompt geven by Michaell Lok of
fyrste voyage of Martine Frobiser
for dyscoverye of Catay &c by northwest

PRO, E164/35. Published with permission of the Public Record Office (U.K.).

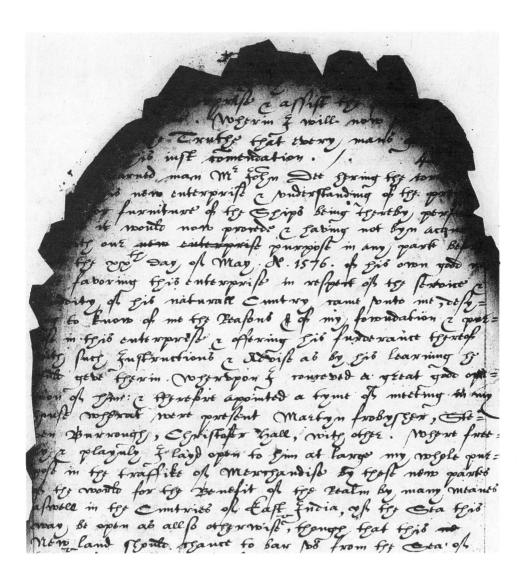

Figure 3: Michael Lok describes the procurement of the first Frobisher voyage (MS 2, Table 1). The manuscript was badly singed in the fire of Ashburnham House, Westminster (23 October 1731), which destroyed or badly damaged 212 volumes of manuscript (out of 958) in the library of Sir Robert Bruce Cotton. *BL*, Cotton, Otho E8/8: f.44R. Published with permission of the British Library.

this collection. Collinson and Stefansson & McCaskill have been the foundation for numerous articles. In the authors' opinion, the best synthesis of the Collinson collection, along with Acts of the Privy Council, has been made by Manhart (1924), but there are many other compilations and syntheses, some predating Collinson.[11]

Finally, there are the manuscripts. Those that have been particularly useful are listed chronologically in Table 1. Special mention should be made of the audited accounts: manuscripts 1, 5 and 6. Manuscripts 1 and 6 are present as two large volumes in the Public Record Office, each bound in leather. They have been summarized (Cooper 1833) and copied (*BL*, Additional MS 39852, copy of about 1824). Parts of manuscripts 1 and 5 have been published (Quinn *et al.* 1979: 193-200; Parks 1935) and all three manuscripts have been transcribed in the thesis of McDermott (1984); the title page of manuscript 1 is reproduced as Figure 2. These accounts contain a wealth of detail as itemized expenses of the Frobisher enterprise, from its inception in 1576 until its final demise in 1581. They are indispensable to any researcher studying the Frobisher ores or voyages.

Manuscript 3 *The Doynges of Captayne Furbusher*, possibly written by a scribe for Michael Lok and with marginalia in Lok's hand, exists as a contemporary copy in the State Papers (*PRO*, SP 12/131/20). The latter was transcribed in 1915 for the Public Archives (now National Archives) of Canada (*NAC*, MG 40/D8/SP12). It contains many new data, but has been previously ignored by researchers, possibly due to a scathing (and perhaps exaggerated) attack on Frobisher.

Manuscripts 7 and 10 are both logbooks of the third voyage. Manuscript 7 is particularly interesting: it is a detailed day-to-day record. It can be divided into four parts, which run consecutively: equipment and provision lists, the outgoing voyage, the sojourn in Baffin Island, and return voyage.[12] Interleaved with the equipment and provision lists is a one-page prayer by the clergyman, the Reverend Robert Wolfall, offering thankgiving for preservation of the ships' company and beseeching their continued deliverance (Kenyon 1980).[13] The log of Christopher Hall is a comparatively factual and terse, daily journal. It ends abruptly at Portsmouth, September 28, 1578.

Why are these voyages so well documented? According to Cooke (1964: 24) "we know something about the financial arrangement of Frobisher's three voyages because they resulted in litigation. Had the voyages been profitable, no doubt these records would have been destroyed". This is partly the answer, but we must also keep in mind that the 'venturers' included in their ranks some very important people: the Lord High Treasurer, the Lord High Admiral, the Earl of Leicester, Secretary Walsingham and many others. The list reads like the Who's Who of 1578. At the very head, and by far the largest shareholder of all, was Queen Elizabeth herself. It was important to preserve all papers for legal actions that lasted into the seventeenth century.[14]

By the nineteenth century, the Frobisher expeditions were all but forgotten. It was the enthusiastic American, Charles Francis Hall, who awakened public interest in this lost enclave of history by accidentally discovering three Frobisher sites (Hall 1864). But, after that, interest again waned. The Frobisher story was revived in the second half of the present century. It then became apparent that Kodlunarn Island (also known as Qallanaaq), the centre of Frobisher's mining activity, was a national treasure that could be endangered by indiscriminate collecting and digging, and in October 1964 it was duly declared a National Historic site. Henceforth, scientific research on the island would require special permission. In 1974 the island was studied by Dr Walter Kenyon of the Royal Ontario Museum, as an archaeological project commemorating the 400th anniversary of Frobisher's exploration (Kenyon 1975a, b). Activity intensified in the 1980s and 90s. In 1981, 1990, 1991 and 1992 multidisciplinary expeditions led by Dr William Fitzhugh of the Smithsonian Institution investigated the Frobisher sites (Fitzhugh 1993a, b, c) and in 1990 and 1991 Drs Robert McGhee and James Tuck assessed the archaeological resources of Kodlunarn for the Canadian Museum of Nature (McGhee & Tuck 1993a; Tuck *et al.* 1993).

By 1990 it became apparent that there was a pressing need to shape and coordinate research. Accordingly, the Meta Incognita Project Steering Committee, an inter-institutional committee funded by research institutes, governments and private sources, was set up to encourage and direct research on the Frobisher sites. Their first meeting was held in November 1990.

Two recent, multi-authored books, one edited by Fitzhugh and Olin (1993), the other edited by Alsford (1993), are largely the product of research sanctioned by the steering committee. Fitzhugh and Olin's book deals primarily with research of the Smithsonian Institution, some predating the foundation of the committee (expeditions of 1981 and 1990), but most made after its creation (expeditions of 1991 and 1992). It covers history, ethnology, historical archaeology, ceramics, geology, mining, botany (dendrography), C-14 dating and metallurgy. Alsford's book spans a similar spectrum but, being the official voice of the steering committee, it is focused more on current research.

The present volume had its beginning in 1988, spawned as a project undertaken at the Natural History Museum (London) and later reared through encouragement of members of the Smithsonian Institution and the Meta Incognita Project Steering Committee. One chapter (here Chapt. 5) was published in an early form in Alsford's book, (Hogarth *et al.* 1993) but is here updated to include data gathered and processed from mid 1991 to early 1993.

Mining during the reign of Elizabeth

Frobisher sailed to Baffin when mining in the British Isles was in its infancy. Methods of extraction of ore were laborious and mines were few. It is true that tin mining had been long established, but individual mines were small and scattered throughout Cornwall and Devon. From 1555 to 1565 Burchard Kranich and William Carnsew the elder extracted a few hundred ounces of silver from lead ore near St Columb Major and Perranporth, Cornwall.[15] Copper mines in Cumbria and lead mines in Yorkshire were worked on a small scale through patents granted in 1568 (Donald 1950, 1951, 1955, 1961). At Clonmines, Co. Wexford, Ireland, an unsuccessful attempt was made by Robert Record, of All Souls College, Oxford, to work a silver mine in 1551 (Donald 1961: 11). In the Leadhills district, Scotland, silver-gold-lead mines were worked, more-or-less continuously, from 1570 to 1592 (Porteous 1876; Cochran-Patrick 1878). The coal trade began in Newcastle in the thirteenth century. The iron works of the Forest of Dean supplied Frobisher with seven miners in 1577.[16] Technology was largely borrowed from Germany and experienced German miners were a major work force in all the above-mentioned regions, except Cornwall (tin mines), Devon, Forest of Dean, and Newcastle.

With the exception of a few Cornish tin mines, operations were conducted as shallow surface grubbings in rather soft rock. Certainly, the comparatively hard and tough rock that Frobisher's men were to encounter, was not within experience. Elizabethan miners were ill-equipped to tackle rocks harder than limestone, shale and slate, breaking with pick, sledge and chisel.

Three widely used, sixteenth century mining methods, that could have been employed in Baffin, deserve special attention, *viz.* pick-and-hammer, plug-and-feather, and fire-setting. For 'pick-and-hammer', the head of a single-tined pick, i.e. one with a hammer at one end of the head, a point at the other, and free from curvature, was held against the rock as a chisel in one hand, and driven home with a hammer in the other (Fig. 4). In Germany, heads, *bergeisen* [mine irons], 20 cm long "were in daily use among the miners", and heads, *ritzeisen* [splitting irons], 45 cm long "were used to shatter the hardest veins in such a way that they crack open" (Agricola 1561: *index secundus-ferramenta*; Hoover & Hoover 1950: 150). Heads were attached to helves but could easily by removed. They needed frequent resharpening and this, no doubt, imposed a heavy demand on blacksmiths (who were present in both the second and third Frobisher voyages). At a silver mine in Lorraine worked in 1530 (Brugerolles *et al.* 1992), the red-hot ends of *bergeisen* were hardened in cool water, but when the tempered tip was used up in moiling, the head was abandoned. During a shift, the sharp point of a *bergeisen* lasted about one hour, after which the head was either resharpened or discarded.

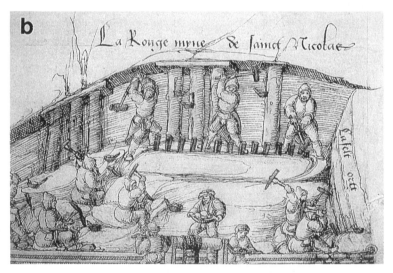

Figure 4: Mining scenes, sixteenth century. a) Saxony; the tools probably resemble those used by Frobisher's miners: *bergeisen* driven by hammers, single-tined picks (also hammer driven), square-sectioned crowbars, and spades with fixed handles; from Lonicer (1551; f. 320R). b) Lorraine; plug-and-feather (feathers not evident) and pick-and-hammer (here two *bergeisen*) operations; from album of H. Groff. Published with permission of the British Library and École Nationale Supérieure des Beaux-arts, respectively.

The 'plug-and-feather' method was especially useful in splitting rock with a well defined grain or fracture pattern. Two facing iron plates or feathers (German *federn*) were inserted in a hole or crevice, possibly formed with picks and small iron wedges or plugs (German *plötz*) were driven between them. If several sets of feathers were inserted parallel to the grain and the plugs in turn driven down with a hammer, the rock was finally rent (G. Agricola in Hoover & Hoover 1950: 118).

For extremely hard, brittle rocks 'fire-setting' was often found useful. When a hot, fired surface was quenched with a dash of cold water, the rock was commonly shattered instantaneously (G. Agricola in Hoover & Hoover 1950: 118-119). This method was particularly applicable to a vein or rock containing abundant quartz, which is notably non-elastic and undergoes phase and volume changes at $573\,^\circ$C; it would be less effective with the more flexible minerals of the 'black ores'.

Rocks were not blasted with gunpowder until the seventeenth century. This development awaited the discovery of an efficient means of boring to produce a cavity to confine the explosive. Gunpower was introduced in Saxon mines in 1627 (Hoover & Hoover 1950: 119 n). Its restricted use at Coniston, Lake District, appears to have followed soon after (Shaw 1983: 20), but in England, even at mid-century it was regarded as somewhat of a novelty (Moray 1665).

Mineralogy as a science was yet unborn. It would seem that 'Agricola' (George Bauer 1494-1555) the German dean of mining technology, had grasped the basic principles of determinative mineralogy; that is, he distinguished minerals according to such properties as hardness, taste, specific gravity and colour, and was aware of certain chemical features. But the methodological assembling of minerals according to chemical classification had to await the advance of chemistry for the systematization of Wallerius, in the middle of the eighteenth century. Assay technology was well developed by 1575 and will be discussed in Chapter 4 and Appendix I.

Into this milieu of miners, assayers, millmen and goldsmiths came Martin Frobisher. Experienced men were few and hard to find, and he recruited largely from the unqualified (probably through the expediency of time): 'miners' were Cornishmen of varying profession, assayers were mainly London goldsmiths with little experience beyond jewelry, most millmen had to be brought from Germany.

The principals

Players in our story range from Privy Councillor to pirate. Frobisher attracted a strange assortment of friends: the courtiers - wealthy, accomplished, adroit, the mariners - by and large and impoverished, illiterate and rough lot. Michael Lok's associates were mostly   London   merchants,   many   well   educated   and   capable   of   assisting   in administration (e.g. Allen, Castelin, Sellman). The assayers, however, may have been ill-chosen: some believed in the magical transmutation of base metals into gold (Agnello, Kranich), others were ambitious goldsmiths (Denham, Humphrey) or German metallurgists with questionable credentials (Bona, Shutz). Then there were those that watched with interest from the sidelines: the editors (Anania, Hakluyt, Pithou), the academics (Dee, Sidney, Vere), spy masters and counter spy (Castelnau, Mendoza, Walsingham), and provisioners and expediters (Baker, Borough, Cole). All in all a group with diverse talents and interests.

These were the men that shaped and drove the enterprise. They were products of the age - rugged individualists.   The following biographical sketches emphasize the broad spectrum of the major players. Most have been mentioned already or will appear in the pages following.

Dramatis personae

Agnello, Giovanni Battista (fl.1581). Venetian assayer residing in London; author of book on alchemy published in London, 1566; tested English currency 1569. Produced gold from rock retrieved from 1st Frobisher expedition; contracted with the Cathay Company until at least 1578. *DBI* **1**: 431-2.

Aldaye, James (b. *c*.1516 - *c*.1585). Proponent of, and perhaps participant in, Thomas Wyndham's voyage to Barbary (1551); merchant in Lapland (*c*.1573); led two barks to southern Greenland in royal (Danish) expedition (1579). With ships' company in Frobisher's first voyage, probably as sailor (1576).

Allen, Thomas (d.1592). London merchant, importer and manufacturer. Supplied cordage to English navy from Danzig (from at least 1567) and from his factory in Woolwich (beginning *c*.1572). Charter member of Muscovy Company (1556); imported merchandise from Baltic (from 1581). Leased his ship *Thomas Allen* (160 t) to Cathay Company in 3rd Frobisher voyage. Pledged and paid £200 to the Company; treasurer (1578-9); commissioner (1579).

Anania, Giovanni Lorenzo d' (*c*.1545-*c*.1608). b. upper Calabria; studied at Naples; returned Calabria (1576); published works on theology, cosmography and geography; author of account (1582) of 2nd and 3rd Frobisher voyages. *DBI* **3**: 19.

Baker, Matthew (d.1613). Foremost Elizabethan shipwright; resident of Deptford. Repaired, maintained and augmented royal fleet (from 1566); reinforced breakwater and deepened Dover Harbour (1583); renovated navy and fortified Tilbury and Gravesend in preparation for Armada (1587-8). For Frobisher voyages, built *Judith* (for William Borough, later sold to Cathay Company), *Gabriel* and six pinnaces.

Best, George (d.1584). Servant of Christopher Hatton (q.v.); professional seaman. In 2nd Frobisher voyage 'gentleman' on *Ayde*, in 3rd, captain of *Anne Francis* and lieutenant of fleet; author of account of all 3 voyages. *DNB* **4**: 415-6.

Bona [Bonner?], Gregory (fl. 1578). German mining technologist. In 2nd Frobisher voyage aboard *Ayde* as assayer and prospector, paid large wage (£51 2s, possibly including travel and relocation expenses), co-discoverer (with Jonas Shutz) of elusive rich 'red ore' in Baffin; in 3rd voyage in *Ayde* as prospector, paid £14 5s (more than twice a miner's wage), scheduled to return with the ships in the fall.

Borough, William (1536-99). English author, navigator, and agent of Muscovy Company. Held commands in several voyages to Fennoscandia and northern Russia (1556-74). With Drake in *Golden Lion* as vice-admiral of fleet in raid on Cadiz (1587). Captain of *Bonavolia* in battles against Armada (1588). Appointed Controller of the Navy (1589). Member of the Levant Company (from 1593). Recruited seamen for 1st Frobisher voyage; pledged £140 to the enterprise; treasurer Cathay Company (1579). *DNB* **5**: 404-6.

Burcot, Dr., see **Kranich, Burchard**

Burde, William, the elder, (1527-86). Customer of London; shareholder and treasurer of Company of Mines Royal (from 1569); employed Robert Denham (q.v.) to manage mines in Cornwall and Wales (1583-6). In 1576 he was one of four who initially sponsored the Frobisher enterprise with £100.[17] Donald (1955: 63-6).

Burghley, Lord see **Cecil, William**

Caesar, Sir Julius (1558-1636). Michael Lok's stepson and son of Caesar Adelmare (court physician, 1558-69); MA Oxon. (1578); appointed judge of High Court of Admiralty (1584), governor of the Company of Mineral and Battery Works (1593), Chancellor of the Exchequer (1606), Master of Rolls (1614); knighted (1603). Pledged £100 to the Frobisher enterprise. *DNB* **8**: 204-7; Lodge (1827).

Carew, Henry (b. *c.* 1525, fl. 1580). Second son of John Carew, landowner of Haccombe, Devon. Godson of King Philip II of Spain; professional seaman; sided with Spaniards before and after voyages. In Frobisher's 2nd voyage, 'gentleman' aboard *Ayde*, and supervised unloading, weighing, and storage of ore at Bristol; in 3rd voyage, captain of *Hopewell* (the 'rear admiral'), and appointed (with 4 others) to Frobisher's advisory council.

Carnsew, William, the elder (d.1588). Diarist and landowner in Cornwall; mined silver deposits in Cornwall (1555-*c*.1562) with Burchard Kranich (q.v.) and the same silver deposits (1583-7) with Robert Denham (q.v.).

Casimir, Duke John [or John-Casimir, Duke], Count Palatin (1543-92). Twice (1568, 1572) led German troops into France to help Protestants. Transported by Frobisher to Flushing (Netherlands) and launched an unsuccessful campaign in Holland (1579); sympathized with Frobisher's problems at Dartford. *NBG* **25**: 525-30.

Castelin [or Castelyn], Edward (d.1585). London merchant and early sponsor of voyages to Guinea (1553-64); joint-owner of 240 ton bark *Primrose* (1557-63). Charter member of Merchant-adventurers of Guinea (1562) and Company of Mineral and Battery Works (1568); Treasurer (1568). Director of Frobisher's works at Dartford (1578-9).

Castelnau, Michel de, Seigneur de Mauvissiere and de Concressant (*c.* 1520-92). French envoy to Scotland (1560); fought in Catholic ranks in France (1567); French ambassador to England (1575-85), where he informed the French king of Frobisher's progress in voyages and furnaces. *BUAM* **7**: 527-8.

Cecil, William, first Lord Burghley (1520-98). English Under-secretary of State (1550-3); Chief Secretary of State and advisor to the Queen (1558-72); Lord High Treasurer and Privy Councillor (1572-98). Pledged £400 in the Frobisher enterprise.[18] *DNB* **9**: 406-13.

Chancellor, Nicholas (*c.*1532-82). Son of Richard (who explored Russian Lapland). Apprentice, then clerk and merchant with the Muscovy Company (1556-76). With Edward Fenton in voyage to Africa, where he died (in Sierra Leone, 1582). Purser in all three Frobisher voyages: in *Gabriel* (1576), *Ayde* (1577), *Judith* (1578); expediter for Cathay Company in interim. Volunteered to overwinter on Kodlunarn Island with Fenton (1578).

Clinton, Edward Feinnes de, Earl of Lincoln (1512-85). English administrator; saw naval action in France and Scotland (1544-7); governor of Boulogne (1547-50); appointed Lord High Admiral and Privy Councillor (1550-5, 1558-85). Pledged £400 to the Frobisher enterprise. *DNB* **11**: 91-3.

Cole, Humphrey (fl.1580). Employed at Royal Mint since 1568; prospector, engraver and instrument maker; commissioner of the Company of Mineral and Battery works (from 1569). Provided instruments for Frobisher's 1st voyage and sulphide additive for furnaces at Dartford. A Humphrey Cole [his son?], 'a learned man of Oxford', was 'preacher' with the embassy of Sir Jerome Bowes in Russia (1583-4). *DNB* **11**: 270.

Courtenay [or Courtney], Thomas (b.c.1540). Third son of Sir William by second marriage; mariner and trader in Devon. Captured ships which resulted in dispersing the first occupants of Fort Dún-an-Óir, Ireland (1579). Captain and owner of *Armonell* of Exmouth in Frobisher's 3rd voyage.

Dee, John (1527-1608). London academic; studied and wrote on chemistry, alchemy, the occult, cosmology, mathematics, cipher and geography. Instructed Frobisher in navigation (1574-5); pledged £100 to his enterprise; commissioner of the Cathay Company (1578- 80).[19] *DNB* **14**: 271-9.

Denham, Robert (d.1605). London goldsmith and metallurgist; director of operations at lead-copper-silver mines in Cornwall and Wales (1583-7). Aboard *Ayde* to Baffin Island as assistant assayer in Frobisher's 2nd voyage, and as chief assayer in the 3rd voyage; assayer for Frobisher in London and Dartford (1577-9).

Desmond, Earl of, see **Fitzgerald, Gerald.**

Diar [or Dyer], Andrew (c.1550-84). Professional seaman; recruited English force sent to Holland (1583); shipped to Tripoli (1584), captured by Turks and hanged. Aboard *Ayde* as pilot in Frobisher's 2nd voyage, and aboard *Hopewell* as master and pilot in 3rd.

Dudley, Ambrose, Earl of Warwick (1528-90). Landowner; distinguished in the siege of St. Quentin (1557) and the battle of Le Havre (1563); created Earl of Warwick (1561); m. Anne Russell (1565) (q.v.). Pledged £400 to the Frobisher enterprise. *DNB* **16**: 97-8.

Dudley, Robert, Earl of Leicester (c.1532-88). Courtier and administrator; appointed Privy Councillor (1559), Earl of Leicester (1564), commander of Protestant forces in Holland (1585- 7). Pledged £600 to the Frobisher enterprise; Leicester's Point and Island (both unidentifiable today) were named after him. *DNB* **16**: 112-22.

Ellis, Thomas (fl.1578). A Thomas Ellis, of John Hawkins' company in the West Indies, was sentenced by the Spanish Inquisition to serve from 1571 to 1578 in the galleys. Sailor in the 3rd Frobisher voyage but not identifiable from paylists; author of an account of the 3rd voyage.

Fairweather, Richard, the younger (fl.1587). English sailor; shipped with Edward Fenton (q.v.) to Africa and South America in 1582 and took up residence on Rio Plate. Master of *Beare Leicester* (owned by his father and Michael Lok) on Frobisher's 3rd voyage.

Fenton, Edward (d.1603). Served in Ireland (1566, 1579-80), led voyage to Africa and South America (1582-3) and commanded *Mary Rose* against Armada (1588). In

Frobisher's 3rd voyage he captained the *Judith* and was Lieutenant General of the fleet. *DNB* **18**: 320-2.

Fitzgerald, Gerald, fifteenth Earl of Desmond (d.1583). Owner of large estate in Munster; fought with English during the Irish uprising of 1579 and with rebels the next year. Accommodated the crew of the *Emanuel*, wrecked on return from Frobisher's 3rd voyage, and helped unload ore. *DNB* **19**: 120-3.

Frobisher, Sir Martin (1535-1594). Seaman (since 1554); enduring interest in Northwest Passage (since *c*.1560); commanded *Triumph* in the Armada battle (1588); knighted (1588); died of wounds received in the battle of Brest (1594). Led three voyages to Baffin Island and pledged £400 to the enterprise. *DNB* **20**: 281-4.

Gresham, Sir Thomas (1519?-79). Statesman who overhauled financial system of England, as Kings Mercer (1551-74) and financial advisor to Queen (1574-9); founded Royal Exchange (1571) and Gresham College (1575). Pledged £800 to the Frobisher enterprise; appointed commissioner (1579). *DNB* **23**: 142-53.

Hakluyt, Richard (*c*. 1552-1616). Scholar and cleric. B.A. (1574), M.A. (1577), Christ Church, Oxford. Chaplain, English Embassy, Paris (1583-89), archaeacon, Westminster (1603-4); rector, Gedney, Lincolnshire (1612-16). Keen interest in Arctic voyages. Published *Principall Navigations* (edns 1589, 1598-1600) and other works. Encouraged and chronicled Frobisher voyages but did not invest. *DNB* **24**: 11-12.

Hall, Charles Francis (1821-71). American journalist and explorer. Engraver (1849-57), newspaper publisher (1858-60), Cincinnati. Led expeditions to Baffin Island (1860-62), King William Island (1864-69), Greenland (1870-71). Published *Life with the Esquimaux* (1864). Pinpointed Frobisher's first mine and stockpiles of coal, SE Baffin Island area (1861-62). *DAB* **4(2)**: 120-21.

Hall, Christopher (fl.1584). English seaman; master of *Tomasin* of Muscovy Company, trading Russian and English merchandise (1581); sailed with Fenton to Africa and South America (1582). In Frobisher's 1st voyage was master of *Gabriel*; in 2nd, master of *Ayde*; in 3rd, aboard *Ayde* as chief pilot of fleet.

Hatton, Sir Christopher (1540-91). English courtier and wealthy landowner; appointed Captain of Queen's Bodyguard (1572), Privy Councillor (1578); Lord Chancellor (1587). Censored narratives prior to publication, of the three voyages of Frobisher. *DNB* **25**: 159-62.

Hechstetter, Daniel (1525-81). Mine developer. Apprentice and mining engineer at copper-silver mines in Rauris Valley nr Salzburg (1541-56); granted privilege to prospect for metals in England (1563), drafted agreement with Haug, Langnauer & Co. of Augsburg to develop mines in Cumberland, where he immigrated (1564); managed copper-lead-silver properties in Lake district for Company of Mines Royal (1568-81). Supplied additive for Frobisher's furnaces (1578-80). *NDB* **9**: 304, Hammersley (1988: 28-31).

Herbert, Henry, second Earl of Pembroke (1534?-1601). Created Knight of the Bath (1553), Earl of Pembroke (1570); appointed Gentleman of the Chamber to King Philip (1554), President of Wales (1587); m., 3rd, Mary Sidney, q.v. (1577). Pledged £600 to the Frobisher enterprise. *DNB* **26**: 189-90.

Humphrey, William, the younger (d.1605). London goldsmith; son of William Humphrey, Assaymaster in the Mint at the Tower (from 1561) and charter assistant to The Company of Mineral and Battery Works (1568). In *Ayde* on Frobisher's 3rd voyage and assayer in Baffin Island; assisted in assaying Frobisher's ore in the Dartford furnaces (1578-9).

Jackman, Charles (d.1581). Professional seaman from Poplar near London. Joint leader (with Arthur Pet) of voyage of exploration for Northeast Passage (1580-1) as captain of barque *William*; lost on return. In 2nd Frobisher voyage was master's mate in *Ayde*, in 3rd master in *Judith*; member of Frobisher's advisory council and volunteered to overwinter in Baffin Island with Fenton's company (1578). Jackman's Sound, Baffin Island, is his memorial.

Kendall, Abraham (d.1596). Professional seaman, distinguished navigator and 'rare scholler'. Master in Drake's voyage to Roanoke (1585-6) and Chidley's voyage to Trinidad (1589); master and pilot in Dudley's voyage to the West Indies and Orinoco (1594-5) and Drake and Hawkins' voyage to Panama (1595-6). Wrote a ruttier [sailing directions] to West Indies and Orinoco, translated by Warner (1899: 80-92). Possibly the Captain Kendall of the *Dennis*, which sank in Frobisher's 3rd voyage (all hands saved).

Ketel, Cornelis (1548-1616). Dutch artist, in Paris and Fontainbleau (1567-72), England (1576-81); returned to Holland (1581); famous for military tableaux and portraits. After 1st Frobisher voyage, commissioned for four portraits of an Inuk brought to England; after 2nd voyage, painted ship *Gabriel* and 7 portraits of 2 other Inuit; also portraits of C. Hall, G. Best, E. Fenton and M. Frobisher. *BUAM* **22**: 331-3.

Kranich, Burchard [alias Dr Burcot] (d.1578). German immigrant metallurgist and physician; worked lead ores in Derbyshire (1552- 5) and silver ores in Cornwall (1555-*c.* 1565). Assayer and metallurgist at London and Dartford with the Frobisher enterprise (1577-8). Donald (1950, 1951).

Leicester, Earl of, see **Dudley, Robert**

Lok, Michael (1532-*c*.1615). Merchant and experienced seafarer "having travelled through almost all the countries of Christianity" during twenty-four years; with Muscovy Company as London agent (1571-6) and treasurer (1576-7). Treasurer of Cathay Company (1577-8) and financially ruined as result of participation in the Frobisher enterprise. *DNB* **34**: 92-3, Williamson (1914).

Lord High Admiral see **Clinton, Edward Feinnes de**

Lord High Chamberlain see **Radcliffe, Thomas**

Lord High Treasurer see **Cecil, William**

Mendoza, Don Bernardino de, Compte de Cluni (1540?-1604). Spanish soldier and diplomat; Chevalier of the Order of Saint Jacques (1572); Spanish ambassador to London (1578-84). Sent reports of the progress of Frobisher voyages, typical specimens of his black ore, and a map, to King Philip II (1578-9). *OCSL*: 382.

Newton, Richard (*c*.1531-87). Bridgwater (Somerset) mariner, trader and shipowner. In the 3rd Frobisher voyage he captained his *Emanuel*, which was wrecked off the coast of Ireland on the return voyage, and the cargo of ore seized by Gerald Fitzgerald (q.v.).

Oxford, Earl of see **Vere, Edward de.**

Pelham, Sir William (*c*.1530-87). English soldier (served in Scotland and France to 1562) and administrator; appointed Lieutenant General of Ordnance (1563), Lord Justice of Ireland (1579) but resigned following year. Pledged £200 to the Frobisher enterprise; appointed commissioner (1578-9). *DNB* **44**: 255-7.

Pembroke Earl of see **Herbert, Henry**

Pithou, Nicholas (1524-98). Protestant French barrister and author; after massacre of St. Bartholemew (1572), fled to Geneva, where he lived out his life. Responsible for a French translation and augmented version of Settle's narrative of Frobisher's 2nd voyage. *NBG* **40**: 342.

Radcliffe, Sir Thomas, third Earl of Sussex (1526-83).   English courtier and diplomatist; appointed Lord Deputy of Ireland (1557-60), Earl of Sussex (1557), Lieutenant General of Ireland (1560-4), Lord High Chamberlain (1572); m. Frances Sidney (1555) (q.v.). Pledged £400 to the Frobisher enterprise. *DNB* **47**: 136-45.

Randall, Hugh (d.1597). Seafarer and trader; sometime bailiff, alderman, and mayor of Weymouth, vice-admiral of Dorset, as well as pirates' friend and accomplice. In Frobisher's 3rd voyage led his bark *Salomon* to Baffin Island and returned with ore.

Russell, Lady Anne, Countess of Warwick (d.1604). Daughter of Francis Russell, earl of Bedford; m. Ambrose Dudley, Earl of Warwick (q.v.) 1565. Pledged £200 to the Frobisher enterprise;[20] Countess of Warwick Sound, Island (now Kodlunarn Island) and Mine, were named after her. *DNB* **16**: 98.

Sellman [or Selman], Edward (fl.1580). London merchant; with William Towerson in Guinea (1557-8); sent to Flushing as trade commissioner (1579). In Frobisher's 2nd voyage as merchant in *Ayde*, in 3rd, as merchant, paymaster of miners, and registrar in *Ayde*; his manuscript of the 3rd voyage has been published (Collinson 1867: 290-316; *SM* **2**: 55-79);[21] in charge of Dartford works (from April 1579).

Settle, Dionites [or Dionyse] (fl.1578). In 2nd voyage aboard the *Ayde* as 'soldier', at close of voyage composed an account, published under the patronage of the Earl of Cumberland; a 'James Settel, gentleman' was with Edward Fenton's company in the 3rd voyage.

Shutz [or Schütz], Jonas (fl.1579). Metallurgist and goldsmith, b. Annaberg, Germany but, at the time of the voyages, resident in London. In 2nd Frobisher voyage aboard the *Ayde*; on Kodlunarn Island set up furnaces and, with two assistants, made assays; assayed ore returned to England in furnaces at London and Dartford (1577-9).

Sidney, Lady Frances, Countess of Sussex (d.1589). Daughter of Sir William Sidney; m. Sir Thomas Radcliffe (1555) (q.v.), later Earl of Sussex; founded Lady Frances Sidney-Sussex College, Cambridge. Pledged £140 to the Frobisher enterprise;[20] the Countess of Sussex Mine, Frobisher's largest, was named after her. *DNB* **47**: 143.

Sidney, Mary, Lady Herbert, Countess of Pembroke (1561-1621). Daughter of Sir Henry Sidney; early acquired knowledge of English, French, Italian, Latin and Greek; closely attached to brother Sir Philip Sidney (q.v.), with whom she co-authored poetry and composed independently; patron of authors; m. Henry Herbert (q.v.). Pledged £100 to the Frobisher enterprise.[20] *DNB* **26**: 204-7; Waller (1979).

Sidney, Sir Philip (1554-86). English soldier, courtier, author and poet, distinguished in battles in Holland (1585-6) but mortally wounded at Zutphen; his poetry underwent several editions. Correspondence shows considerable interest in Frobisher voyages; he pledged £200 to the enterprise. *DNB* **52**: 219-34.

Sussex, Countess of, see **Sidney, Lady Frances**.

Sussex, Earl of, see **Radcliffe, Sir Thomas**.

Vaughan, William (d.1580). Tenant miller at Bignores, near Dartford; "one of her maiestes yomen of her chamber"; founder of the Dartford Grammar School. Property acquired by Cathay Company in 1578 as mill and furnace site.

Vere, Edward de, seventeenth Earl of Oxford (1550-1604). Courtier, statesman, poet. Called poetical earl; claimed by some to be author of Shakespeare's works. Inherited earldom (1562); MA Cambridge (1564), Oxford (1566); special commissioner in trials of Mary Queen of Scots (1586) and Philip Howard (1589); fought against Armada (1588); apptd Great Chamberlain and Privy Councillor (1604). Purchased £1000 of Cathay Co. stock from Michael Lok (1578). *DNB* **58**: 225-29.

Walsingham, Sir Francis (1530?-90). Privy Councillor and Secretary of State for England (1573-1590). Shareholder in The Muscovy Company (from 1569). Pledged £800 to the Frobisher enterprise; organized and controlled England's espionage and counter espionage network, which intercepted many letters reporting details of the Frobisher enterprise, which were sent by Mendoza to Philip II. Died insolvent. *DNB* **59**: 231-40.

Ward [or Warde], Luke (d.*c*.1592). Professional seaman. With Muscovy Company (until 1577). Fought against pirates (1578); with Edward Fenton in voyage to Africa and South America as vice-admiral and captain of *Edward Bonaventure* (1582-3). Captain of *Tramontana* in battle against Armada (1588); on Channel patrol (1589-91). In 3rd northwest voyage, Frobisher's confidant and 'gentleman' on *Ayde*. *DNB* **59**: 350.

Warwick, Countess of, see **Russell, Lady Anne**.

Warwick, Earl of, see **Dudley, Ambrose**.

Wiars [or Wyares], Thomas (fl. 1578). English mariner. Author of letter describing ficticious island in north Atlantic, printed by Hakluyt (q.v.) in 1589, with land and surrounding fishing rights sold to Hudson Bay Company in 1675. In 2nd Frobisher voyage, boatswain in *Michael*; in 3rd, passenger in *Emanuel* of Bridgwater.

Williams, William (d.1599). Assay master at the Mint in Ireland, then appointed Deputy Assay Master at the Mint in the Tower (1567); from 1583 Joint Assay Master with son Walter. Assayed first specimen of Frobisher ore (1577) and also made the last assay on record (1583).

Winter [or Wynter], Sir William (d.1589). Career seaman with experience in Scotland, Jersey Islands, the Levant, France and Ireland (1544-80); knighted (1573) fought against

Armada (1588); appointed Master of Ordnance for life (1557). Pledged £300 to the Frobisher enterprise; commissioner (1577-9). *DNB* **62**: 220-2.

Wolfall [or Woolfall], Robert (*c*.1545-1610). Anglican clergyman holding posts in diocese of Bath and Wells. Sailed as clergyman aboard *Judith* in Frobisher's 3rd voyage. Francis (1969).

Yorke [or York], Gilbert (d.1595). Professional seafarer; served in Ireland under Sir John Parrot, as captain (1579), and under William Winter, as vice-admiral (1580); against pirates (1584); with Drake and Hawkins to Panama (1595), where he died of sickness. In Frobisher's 2nd voyage was captain of *Michael*, in 3rd captain of *Thomas Allen*, vice-admiral of fleet and member of Frobisher's council[22]; commissioner (1579-80); York's Sound, Baffin Island, is named after him.

Notes

1. Most biographical data have been taken from George Frobisher's manuscript *the Frobisher story*, 54 p. (1978) and through communication with the authors. Thus we have used 1535 as Martin Frobisher's birth date in both the beginning of this chapter and in the biographies. George Frobisher, who has done extensive genealogical research on the family, asserts that the birth date 1539, given by Eliot (1917), is incorrect.

2. This strange belief was reiterated by Count Bernardino de Mendoza, the Spanish ambassador to London, in 1578: "the assays [of ore returned in the second Frobisher voyage] showed the presence of gold, which it is against reason to believe can be found in such a cold land as that, so far north as it is" (Hume 1894: 583). Philip Sidney writing of the first voyage, reported that [Garrard, who discovered the first specimen or ore], "not believing that precious metals were produced in a region so far north, considered it [the ore] of no value" (Pears 1845: 119, 226).

3. Account of Michael Lok of the first voyage: "Bought of Richard Clarke, goldsmithe 1 tuche stone in case of lether li 0.5.0 [5 shillings]". *PRO*, E 165/35, f. 46. A touchstone is a dark rock used to test the purity of noble metal, with a streak produced by rubbing a sample metal on its surface. A set of needles of known composition are used as standards (Hoover & Hoover 1950: 252-8; Moore & Oddy 1985).

4. The pledges and outstanding commitments are listed in the audited accounts of Thomas Neale and William Baynham (August 12, 1580; May 6 1581) in *PRO* E 164/36, ff 317-23. They have been itemized by Shammas (1971: 199-200) and discussed by Shammas (1975) and McDermott (1984: 42-59).

5. *BL*, Cotton MS, Augustus 1, No. 1. "A brief Remembrance of sundry foreign Regions Discovered". Inscribed by Dr. John Dee, 1577.

6. The French edition is augmented with several paragraphs describing Frobisher's landing and the Inuit captives. Translation from the English edition and augmentation was made by Nicholas Pithou. The Latin and German editions were translated from Pithou's volume (*SM 2*: 226). There are two variants of the English edition as explained by Parks (1938) and Baughman (1938).

7. It is unlikely that the artist was John White, the founder of the second Virginia colony, who does not appear on the attendance sheets (*PRO*, E 164/35, 1577), cf. statement in Hulton (1961: 18) "that he [White] was actually there".

8. It remains a mystery why Best's narrative was ignored in its day. There are no other contemporary editions or translations, as distinct from the books of Settle and Ellis, nor was the narrative included in Hakluyt (1589), see Quinn *et al.* (1965). The book is found in various collections of voyages, starting with Hakluyt (1598-1600), copied in the Old South Leaflets (Best 1900) and paraphrased by Kenyon (1975b).

9. The book is dedicated, in two leaves, to Francesco Bifoli and dated December 17, 1582. It is cut in octavo and bears no pagination. Part 1 includes the iceberg illustration of Ellis, as a foldout, and 26 leaves of text of the second voyage. Part 2 includes the Inuit figures of Settle, as another foldout (reversed motif), and 13 leaves of the third voyage. The illustrations have, therefore, been transposed in this volume (from the New York Public Library). One other example of this book is known (in the Biblioteque Nationale, Paris).

10. A more accessible edition, describing the Frobisher voyages is Hakluyt (1927 **5**: 131-276).

11. A partial list of publications follows. This excludes those named heretofore, but includes some that will reappear later.
Archaeology: Fitzhugh (1993a, b), McGhee & Tuck (1993a) Tuck *et al.* (1993).
Business and finance: Scott (1968: 76-82), Shammas (1975).
Ethnology and Inuit: Cheshire *et al.* (1980), Hall (1864), Hulton (1961), Rowley (1993a, b).
Geography and marine: Becher (1833), Buerger (1938), Gunther (1927), Taylor (1938), Taylor (1956: 206-9).
Geology and mining: Blackadar (1967a), Hogarth (1993a, c), Hogarth *et al.* (1985), Roy (1937).
Popular narratives: Barrow (1818), Boetzkes (1964), Bourne (1868: 124-77), Bruemmer (1966), Caswell (1969), Cooke (1964), Jackson (1993), Jones (1878), Kenyon (1975a), King (1955), Leslie *et al.* (1855: 165-79), McFee (1928), McGhee & Tuck (1993b), Pearson (1966), Powell (1909: 175-87), Richardson (1861: 76-88), Rickard (1947: 99-121), Rundall (1849: 7-32).
Unsolved problems: Babcock (1922), Christy (1897), Ehrenreich (1993) Fitzhugh (1993a), Harbottle *et al.* (1993), Harbottle & Stoenner (1987), Olin (1993), Sayre *et al.* (1982), Unglik (1993), Wayman & Ehrenreich (1993).

12. In MS, *PL* 2133 the following pagination is inked on the upper part of the leaves (right hand on recto, left hand on verso): pp. 5-15 equipment and provision lists, pp. 16-25 the outgoing voyage, pp. 25-61 (July 1-Sept. 3) sojourn in the Baffin Island area (this part published by Kenyon 1981), pp. 62-73 the return voyage. The citation to Pepys Library manuscript is incorrect in Kenyon (1980 & 1981).

13. Wolfall was the principal clergyman on the third voyage, and first bestirred the spirits of the ships' company during the violent storm of July 3, when all hope seemed to be lost. He officiated in the first Anglican communion service and the first thanksgiving service in North America, on Kodlunarn Island, July 22, 1578. Frobisher, who had been requested in his *instructiones* to bring with him a "minister or twoo" (*SM* 2: 160) actually brought three: Besides Wolfall, there was John Ayvie [also spelled 'Ivie'], minister and miner on the *Thomas* of Ipswich and Stephen Ridisdaile, minister on the *Ayde* (*PRO*, E 164/35, ff. 185, 254, 301: *ibid*. 36, ff. 127, 152).

14. McDermott (1984: 21) notes that in 1615 one creditor sued Michael Lok for £200 owing on the accounts of the Frobisher enterprise.

15. Information on this enterprise can be found in Donald (1950). Another important and, hitherto, unpublished manuscript is *CRO*, DDME/821/34 *The silver-lead mines of Cornwall and the activities of Doctor Burchard Cranach* by William Carnsew, Sr., pp. 17, *c*. 1587.

16. *PRO*, E 164/35, f. 111, 1577.

17. The others who proferred £100 in 1576 were Thomas Gresham, Michael Lok and William Bonde.

18. Eight members of the Privy Council pledged a total of £3000.

19. On the basis of misinterpretation of John Lee for John Dee (*PRO*, SP 12/123/50; *SM* 2: 154), Ward (1926) has argued that Dee was actually present in the second voyage. This interpretation of secretarial hand disagrees with that of McDermott (1984) and Hogarth (1993b), who assert that the surname can be nothing but 'Lee', gentleman on the *Ayde*. Note that John Lee was also present on the third voyage (*HL*, HM 715, p. 15) as lieutenant [aboard the *Judith*?] and volunteer "to inhabit the New Land" for a year.

20. Besides the Ladies Anne Russell, Frances Sidney and Mary Sidney, other woman shareholders were the Ladies Anne Talbot (pledged £35), Anne Francis Kyndersley (pledged £210) and Elizabeth Martin (pledged £70).

21. The manuscript (No. 8, Table 1) bears marginalia with important data, omitted from the published versions.

22. The council consisted of Edward Fenton, George Best, Richard Philpott, Henry Carew and Gilbert Yorke (*Instructiones, SM* 2: 158), who were "to consult and confere [with Frobisher on] what is best to be done in weightie causes incident on land". Frobisher was accused of seeking advice of the council far too infrequently and acting autocratically.

CHAPTER 2

The Three Voyages

The three voyages heralded an awakening in northern exploration, both geographical and mineral. For Europe in general, they introduced a new region in the Americas; for England in particular, they suggested a new access to the Orient and the possible occurrence of precious metals in the far north; for gentlemen of the Court, they created the excitement that goes with wild speculation; and for Frobisher, they provided a test for his long-cherished dream, exploring the Northwest Passage.

These expeditions were fielded in three successive years and developed from a purely geographical adventure (first voyage), to intense prospecting (second voyage),[1] to serious mining (third voyage), on a steeply ascending scale of men and materials. The first voyage was small (37 men, 2 ships), the second moderate (145 men, 3 ships), the third immense (at least 397 men,[2] 15 ships).

It is not our intention to recount all the available details of these voyages. The story can be well assembled from information in Stefansson & McCaskill (1938) and Kenyon (1981). Rather, we wish to outline the main events and prepare a background for the theme of this monograph, the mining enterprise of Martin Frobisher.

The first voyage, 1576 [3]

On June 7, Frobisher set sail from Ratcliffe, a small community to the east of London. Bearing in mind that, at this time, the Julian calendar was in use (10 days in advance of ours), and a six-week Atlantic crossing lay ahead, this was late in the season to begin an arctic voyage. At his command were two barks, the *Michael* of 25 tons and the *Gabriel* of 20 tons, and a closed-deck pinnace of 7 tons, vessels no larger than many of today's coastal fishing boats. On board was a total complement of 37 [4] (Table 2).

They steered northward but "contrary" winds delayed their progress and it was June 25 before they came about to the west, just short of the Shetland Islands. Following a violent storm in which the pinnace and four mariners were lost, 'Friseland' [Greenland] was sighted, "rising like pinacles of steeples and all covered with snow", but the land being "so full of yce" they remained off shore. Fog descended and for several days they proceeded blindly, during which time "the *Michael*, mistrusting the matter, conveyed themselves privilie away ... and returned home".

On July 24 they began crossing Davis Strait. Then on July 26 'Cape Labrador' [Resolution Island] came into view, but again ice prevented landing and they steered northwards. Finally the wind dropped and the sun came out; on August 1 their

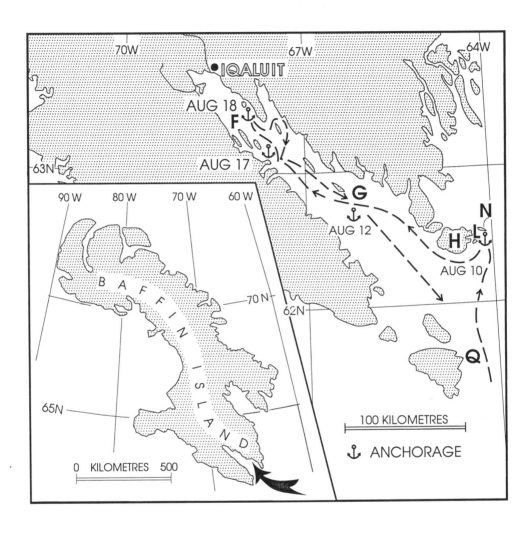

Figure 5: Frobisher's first voyage (1576), showing route in 'Frobisher's Straits' [Frobisher Bay], southeast Baffin Island, and anchorages. Abbreviations: F = Five Mens' Sound, G = Gabriel's Island, H = [Greater] Hall's Island, N = North Forland, L = Little Hall's Island, N = North Forland, Q = Queen's Forland. Iqaluit is a modern name, all others are Elizabethan.

Table 2: *Summary of ships and personnel of the three voyages of Martin Frobisher*

	First voyage (1576). M. Frobisher, general			Second voyage (1577). M. Frobisher, general		
	Michael	*Gabriel*	*Pinnace*	*Ayde*	*Michael*	*Gabriel*
Captain	M. Kindersley	M. Frobisher		M. Frobisher	G. Yorke	E. Fenton
Master	O. Gryffen	C. Hall		C. Hall	J. Beare	W. Smithe
Master gunner		R. Slight		R. Coxe		
Boatswain	P. Whithere	R. Frynd		H. How	T. Wiars	H. Peale(?)
Pilot				A. Diar		
Surgeon		P. Becket		E. Cooley	W. Edwards	
Purser(s)	N. Chancellor			N. Chancellor		J. Netherclyfte
Tonnage	25	20	7	200	25	20
Complement*	13	15	4	117	16	13
'Miners'				8		
Owner(s)	venturers	venturers	venturers	Cathay Co.	Cathay Co.	Cathay Co.

(**table 2**, cont.)

Third voyage (1578). M. Frobisher, general; E. Fenton, lieutenant general; G. Best, E. Harvey, J. Lee, lieutenants

	Company's ships				Commissioned ships			
	Ayde	*Judith*	*Michael*	*Gabriel*	*Thomas Allen*	*Hopewell*	*Anne Francis*	*Thomas* of Ipswich
Captain	M. Frobisher	E. Fenton	M. Kindersley	E. Harvey	G. Yorke	H. Carew	G. Best	W. Tanfield
Master	R. Davis	C. Jackman	B. Bull	T. Price	W. Gibbes		J. Beare	
Master gunner	J. Debois	R. Burnit		W. Lane				
Boatswain	A. Creake		G. Syllebin	J. Jacklin				
Pilot	C. Hall	L. Skypwith		J. Hunt		A. Diar		R. Coxe
Surgeon	J. Harwood	J. Paradice	R. Hind					- Beton
Purser(s)	T. Thorneton J. Cominge	N. Chancellor				R. Chaynalse J. Cara		
Tonnage	200	100	25	20	160	150	130	130
Complement*	134	38	13	22	13	23	20	20
'Miners'	27			7	11	21	17	16
Owner(s)	Cathay Co.	Cathay Co.	Cathay Co.	Cathay Co.	T. Allen	R. How	F. Lee	T. Bonham

(table 2, cont.)

Third voyage (1578), cont.

	Commissioned ships			Non-commissioned ships			
	Francis of Foy	*Mone of Foy*	*Beare Leicester*	*Salomon*	*Emanuel*	*Armonell*	*Bark Dennis*
Captain	H. Moyle	- Upcot	R. Phillpot	H. Randall	R. Newton	T. Courtenay	- Kendall
Master	T. Morrys	J. Lakes	R. Fairweather	- Hudson	J. Smythe	J. Anthony	- Dabnay
Boatswain							
Pilot	W. Luce	S. Beare		J. Lunt			
Surgeon							
Purser(s)		W. Trelegan					
Tonnage	130	100	100	120	100	100	100
Complement*	15	13	30	13	12	14	11
'Miners'	10	10		10	7	12	8
Owner(s)	J. Rashley	W. Mone	R. Fairweather M. Lok	H. Randall	R. Newton	T. Courtenay	N. Dennys

* Minimum numbers from publications and manuscripts

much-sought landing took place. They anchored off Hall's smaller island. Christopher Hall, master of the *Gabriel*, in a small boat with four men, rowed ashore, explored the premises and scaled the highest peak.

The next two weeks saw the *Gabriel* exploring the partially ice-choked Frobisher Bay, mainly along the southern shore (Fig. 5). Notable was an incident which took place August 21, when 5 men and the only boat were taken by the Inuit, leaving the *Gabriel* with "13 men and boyes so tyred and sik ... they were [not able] to proceed ... any further" and worse, they were deprived of their sole means of disembarkation. With heavy heart, Frobisher retraced his journey, arriving back in London October 9. He had discovered an opening in Baffin Island, 'Frobisher Straits', which he believed to be the Northwest Passage, but which we now know was Frobisher Bay. On board was an Inuk captive and a token of possession - an unusual black rock.

As the discovery of this black rock and opinions on its value are pertinent to our theme, they will be considered in some detail. An account, which hitherto has escaped historians, is contained in Lok's *Doynges of Captayne Furbusher*[5] as follows:

> In that first voyadge on a littell Illande of Rock of a half a mile circuit, w[hi]ch they namyd hawlls Illand,[6] after the name of Christofer hawll m[aste]r of the Shippe, who was the first man that Landed theron, Captaine Furbysher not being on land at all, but remayninge still in the Shippe at Sea x milles from yt. On this Rocke was found by chaunce by a Maryn[er] namyd Robert Garrard,[7] who was one of the .v. men w[hi]ch afterward weare taken w[i]th their boate by the people of that Countrye, a blacke stonne, as great as a halfe pennye loaf being on the ground losse, w[hi]ch he thought to be a Seacole, w[hi]ch he brought a boarde the Shippe to prove yf yt woold burne for fyre, whereof they had lacke. This stone beinge thus brought home, Captaine Furbysher gave a pece therof unto Mychael Lok, Sayenge that it was the fyrst thinge that he fownd in the newe Land, and therefore gave yt unto him according to his promyse.

In another account, Michael Lok states:

> in xiii day of October laste, m[aste]r Furbosher gave me a stone, abord his shyp, sayenge, that according to his promesse, he dyd gyve me the fyrst thynge that he found in the new land, w[hi]ch he gave me openly in presens of 2 yonge gentlemen whome I know not; but Rowland York was then in the shyp, and they for the straungeres thereof brake of[f] a pece w[hi]ch they caried awaye with them.[8]

Then, in 1577, Philip Sidney wrote to Hubert Languet as follows (translation by Pears 1845: 118-119, 226):

> Frobisher ... touched at a certain island, and there by chance a young man, one
> of the ships company, picked up a piece of earth, which he saw glittering on the
> ground... and kept the earth ... till his return to London. And there when one
> of his friends perceived it shining in an   extraordinary manner, he made an
> assay.

A fourth version, and the one most commonly quoted, is given by George Best (*SM* 1:
51):

> One [of his company in the first voyage] brought a peece of blacke stone, much
> lyke to a seacole in coloure, whiche by waight seemed to be some kinde of
> mettal or Mynerall ... And it fortuned, a gentlewoman, one of ye adventures
> wives, to have a peece thereof, which by chance she threw and burned in the
> fire, so lo[n]g, that at length being taken forth, and quenched in a little vinegre,
> it glistered with a bright Marquesset of golde.   Whereupon ye matter being
> called in some question, it was brought to certain Goldfinders in London, to
> make assay therof, who indeed found it to hold gold, and that very ritchly for the
> quantity.

What is the true story? Certainly the most credible is the first version. Versions 2 and
3 may be essentially true but are incomplete.   Version 4 is, for the most part,
implausible. But probably all versions contain an element of truth. Common to the first
(Lok's detailed account of the half-penny loaf) and the fourth (involving the fragment
given to the wife of a venturer) is the fact that a portion of the original specimen was
subjected to fire.

What of the ore itself? Best (*SM* 1: 51) describes it as a heavy coal-black stone. Philip
Sidney (Pears 1845: 118-9, 226) stresses its glitter ['resplendentem'] and William Adams
(Fox 1910: 115) its hardness. Subsequently there are many references to a hard, shining,
heavy, black ore in both published and unpublished accounts. The term *black ore*, as
frequently used in contemporary accounts of the voyages, implied a black coarse-grained
rock containing considerable amounts of a mineral, now known as hornblende, occurring
in the Baffin Island area, and believed to contain gold. The term is here extended to
similar rock, *possibly* mined by Frobisher in 1578. Our specimens of 'ore', ultimately
derived from Baffin, were pitch-black, could scratch glass, reflected light off large
hornblende cleavages and were three times as heavy as water, attributes agreeing well
with the contemporary descriptions. As will be shown below (Chapter 5), they belong
to a number of igneous and metamorphic rock types.

Let us now turn to the assays of this first specimen. The goldsmith Giovanni Battista
Agnello, was the cause of all succeeding activity. He began with two qualitative assays
which suggested high gold content and led to a third, in which the value of gold was
given at the rate of 25 ounces per ton, a very high tenor of gold, today as well as

yesteryear. This was followed by at least eight check analyses by other assayers, including William Williams (assaymaster of the mint at the Tower), master Wheeler (Williams' assistant), George Needham (a director of the Mines Royal), Edward Dyer (a courtier with some metallurgical experience), Geoffrey LeBrum (a French assayer residing in London) and Jonas Shutz (a German assayer in London). With the exception of Shutz, all reported that their specimen contained nothing of value and no gold at all. Still, Agnello stuck to his story and convinced Sir William Winter (an early 'venturer' in the enterprise), Sir John Barkley (who had an ear at the Court), Michael Lok (the prime motivator of the voyages) and Frobisher, of the great worth of the ore. During the winter, a company known as the Cathay Company [9] was organized and plans were laid for obtaining more ore. Equipment and supplies were stored in the warehouse of George Winter (Sir William's brother), off Seething Lane and near the Tower, in anticipation of an expedition the following year.[10] The stage was thus set for the succeeding arctic voyages.

The second voyage, 1577 [11]

The second voyage was essentially a mining and prospecting venture and all other objectives were given low priority. The *instructions*[12] (Collinson 1867: 117-120) can be summarized in four paragraphs:

1)   to mine such ore as was available on Little Hall's Island. When the miners were thus occupied, Frobisher was authorized to take one of the barks and look for additional supplies of ore, keeping a watch for the five men lost the previous year. Assays were to be made on-site but grades, tonnages and the purpose of the voyage were to be kept strictly secret.

2)   To establish a temporary colony on 'Friseland' [Greenland] as a prelude to possession. Similarly, "if it bee possible [,to] leave some persons to winter in the Straights" [Baffin Island].

3)   Time permitting, to proceed with the exploration of the Northwest Passage.

4)   To bring to England between three and ten Inuit, but Frobisher was to "cawse [his] companie to doe towards the sayd people as maye give le[a]st cause of offense ... to wynne both friendshippe and lykinge".

Three vessels were appointed for the voyage: the *Ayde* a galliass of 200 tons burden (Fig. 6),[13] and the two barks of the last voyage, the *Michael* and *Gabriel*. Frobisher was authorized to man his vessels with 120 persons "uppon w[hi]ch number [he should] not in any wise exceade".[14]

Departure from Blackwall took place May 26. They stopped at Harwich for four days to take on provisions, but during the interval Frobisher received letters reminding him

Figure 6: The *Ayde* in battle, Smerwick Harbour, 1580. The ship appears to have been rebuilt and enlarged shortly after this episode. It had been Frobisher's 'admiral' in the second and third voyages. Portion of watercolour by William Winter, the younger. *PRO*, MPF 75. Published with permission of the Public Record Office (U.K.).

"not to exceede hys complemente and number appoynted hym, whyche was one hundred and twentye persons: whereuppon he dyscharged many proper men, whiche wyth unwilling myndes departed" (George Best in *SM* 1: 53).

The three ships proceeded up the east coast of England, to the Orkney Islands (where they landed, took on fresh provisions and found a silver mine), explored south Greenland for four days (without landing), and then crossed Davis Strait. On July 17 they "made the North Forlande perfite" (George Best in *SM* 1: 56).[15] After dropping anchor, Frobisher, his three assayers, and others of his company, rowed a pinnace to Little Hall's Island. "They remayned on that Iland almost a hole day, in w[hi]ch tyme they passed over and over all the Illand rownd abowt and crossed everie waye seeking and serching, and could find no mynes at all"[16] nor could a piece of 'black ore' be discovered as "bigge as a walnut". But the next day Frobisher explored Greater Hall's Island and surroundings, where he found several occurrences of rock similar to the original discovery.

Then followed 10 days' exploration of the southern shore of 'Frobisher Straights' [Frobisher Bay]. On July 28 Frobisher took the two barks and crossed over to the north shore. There they dug and stockpiled 20 tons of black ore on 'Leicester's Island' in Beare Sound (precise location not known at present). The next day they anchored off 'Countess of Warwick' [Kodlunarn] Island and mining and loading began in earnest.[17] Later in the month Jonas Shutz discovered his red and yellow ores on 'Jonas Mount', a hill near the head of 'Seate Gulf' [Napoleon Bay].[18] This rock was not loaded but a sample of the red variety was assayed by Burchard Kranich in November 1577 and was recorded as "bearing at the rate of twoo onces in a hundred weight".[19] Naturally this sparked considerable excitement in London.

On August 21 Best reported: (*SM* 1: 75-6):

> And it was now good time to leave, for as the men were well wearied, so their shoes and clothes were well worne, their basket bottoms torne out, their tooles broken, and their shippes reasonably well filled. Some with over-straining themselves received hurtes not a little daungerous, some having their belies broken, and other their legges made lame. And about this time ye ise began to congeale & freese about our ships sides a night.

Then they took down their tents and prepared for departure homeward. The following day anchors were hoisted and they sailed out of the Bay. In late September the *Ayde* and *Gabriel* passed the Isles of Scilly and anchored at Bristol. The *Michael*, which had been separated in a storm, returned to London via Scotland and the east of England. This time they had on board three Inuit and 158 tons of 'black ore' (140.9 tons in the *Ayde* and *Gabriel* and 17.45 tons in the *Michael*).[20]

It is interesting to note the extreme precautions taken to ensure the safety of their cargo. On September 18 the Council proclaimed that the ore be kept temporarily in Bristol Castle, locked firm with three separate locks and three separate keys.[21] Apparently this was not considered sufficiently protective or the directive fully explicit, for on October 17 it was followed by a second ruling:[22] the ore from the *Gabriel* and *Ayde* was to be kept behind a great door of the castle, clamped with "foure sundrie locks and foure several keyes". Sir Richard Barkley, the Mayor of Bristol (Thomas Colstone), Michael Lok and Martin Frobisher were each to be custodian of one key. Thence the following year, the ore was to be transported to London, where it was to be "molten downe and tryed in some fitt and convenient place". Similarly, ore from the *Michael* was to be stored temporarily in the Tower of London, sealed behind a door with four great locks and to be opened with four great keys, of which Frobisher was to have one, the Queen's officers two, and Lok one.

It is also interesting to record the tight security imposed to protect all secret information. Publication was limited to authorized accounts of the voyages; in other words, they could be printed only after rigid censorship. Thus magnetic declinations, bearings and distances were largely deleted from the contemporary accounts of Best and Settle. Furthermore the published maps were drawn so roughly that for fully 200 years geographers placed Frobisher's landfalls on Greenland. However the information ban was almost eluded by one Typottes, a notary of London, who had his manuscript, the Spanish translation and all copies of his book seized from the press and confiscated by Sir William Winter.[23]

We may now reflect on the accomplishments of Frobisher in this voyage. He had loaded a good tonnage, considering the tools available, the short summer season, and a small group of miners inexperienced in hard-rock quarrying. It appears that in this voyage rather few on-site assays were made and the ore was quickly loaded with a view to later testing in London.

This concludes the positive side of the ledger. True, Frobisher had returned with three Inuit, but they were captives taken at the expense of incurring the wrath of the local inhabitants. Furthermore, his company had violently bestirred them on one occasion, and on another had precipitated a skirmish in which five or six of the natives were slain (Fig. 7).[24] After these encounters the Inuit kept a discreet distance from the voyagers.

No attempt was made to colonize Greenland or Baffin, both areas that Frobisher regarded as inhospitable. In exploration, his critics could claim one more failure: he did not proceed westward beyond Countess of Warwick Sound. This was, of course, due to his preoccupation with the main goals of the expedition - the mining and loading of ore.

Figure 7: Skirmish between Frobisher's men and Inuit (1577), sketched by an unknown artist. View from Kodlunarn Island looking northward into Diana Bay. The rock at middle-right is biotite gneiss with characteristic steep westerly dip. Copy in the John White album, British Museum. Published with permission of the Trustees of the British Museum.

That Frobisher was fiercely independent, a poor accountant, and inconsiderate of economy, is apparent in the Exchequer papers. That is why the Council felt obliged to inform him, on two occasions, of the maximum size of the expedition. Even with that reminder, Frobisher exceeded his complement by at least 23 men. This strained the provisions to the limit and was said to be responsible for an early return of the ships. In addition, he had on the payroll 19 mariners (which probably included convicts) and 6 miners who were found "unfyte for the voyage" and discharged beforehand.[25] The net result was an additional charge against the Company of £400.[26]

The third voyage, 1578

Preparations and beginning

Hardly had Frobisher's ships returned from the second voyage than the Cathay Company was planning a third. In the new assay furnace built at London, Jonas Shutz reported that a ton of 'black ore' from Kodlunarn showed a gross value of £40. Burchard Kranich claimed the 'red ore' from the locality later known as Napoleon Bay was three times as rich, and, should a third voyage be fielded, Frobisher promised to load 500 tons.[27] London buzzed with rumours of a new Peru and feverish preparations were made to mount a grand expedition the following spring. In January the Council appointed new commissioners to look after details.

Frobisher's return from the second voyage had not gone unheeded by other nations. The French ambassador, de Castelnau, informed King Henry III of France and Catherine de Medici that 'Forbichet' had claimed new territories for England "pleines de mines d'or" (Read 1926). Frobisher was said to have intercepted a letter in which the King planned to secretly arm 12 ships and take possession of the mines and surrounding country.[28]

More detailed correspondence took place between the Spanish ambassador to England, Count Bernardino de Mendoza, and King Philip of Spain. Mendoza reported that in the second voyage a trench was dug three fathoms deep, from which Frobisher was said to have received 250 tons of ore. The King requested a detailed map of 'Frobisher Straights', specimens of typical ore and a spy to be placed on the third expedition. In April, 1578, Mendoza smuggled to Philip two specimens of 'black ore' and one of 'marcasite' which, after assay in Spain, were pronounced worthless. Delivery of the map was delayed because of fear of discovery and reprisals, but the agent was duly placed in one of the ships and made his report after disembarkation. Additional specimens were requested, but it was well after return of the third voyage when they and the marine chart were dispatched. Naturally the specimens held nothing of value, but the King thanked Mendoza warmly, for his excellent map (Hume 1894: 567-9, *et seq.*).

Russia also entered the arena. She forwarded to London an 'advertycement', what would be called a stiff diplomatic note today. From the Yugorskii area (near the Ob delta and 5000 km east of Baffin) came word from a royal rent-gatherer of Russia: in this very region Frobisher had captured a native man, woman and child, but he had also abandoned five of his own men. The inhabitants, called Paky Samwedey, had complained bitterly.[29]

Now, for the third voyage Frobisher acquired 15 ships (Table 2). By this time the Cathay Company had its own four ships, *viz*. the *Ayde*, *Judith*, *Michael*, and *Gabriel*, but Frobisher was awarded patent to press into service additional vessels as well as miners, mariners and other men useful to the voyage.[30] When the *instructiones* were finally penned, Frobisher was authorized to hire six additional vessels only, viz. the *Thomas Allen*, *Hopewell*, *Anne Francis*, *Thomas* of Ipswich, *Francis* of Foy and *Mone* of Foy. McDermott (1984: 103) has pointed out that the *Beare Leicester*, jointly owned by Richard Fairweather, the elder, and Michael Lok, does not appear in the listing and rightfully belongs with the non-commissioned ships, although Lok later arranged for it to have the same privileges as those that were commissioned, and she is normally included (as in Table 2) with the commissioned vessels.[31]

The formal indentures, or charter parties, were inscribed on parchment by a professional penman, and copies delivered to the ships at Harwich on the outgoing voyage. According to a specific clause, payment to each owner of a commissioned ship was to be calculated at the rate of £5 2s 8d the ton of ore delivered in England but, by agreement, Frobisher's non-commissioned ships were admitted under the same clause.[32] Richard Newton, owner of the ill-fated *Emanuel*, was granted the ore landed in Ireland (see Chapter 3) but Nicholas Dennys, owner of the *Bark Dennis*, received nothing for the loss of his ship.

The expedition was large by any standard, perhaps the largest in the annals of Arctic voyages up to the present century. Aboard the 15 vessels we know the names of 397 officers and men (Table 2), but the list is not complete, as the complement of the non-commissioned ships is not given in the Exchequer papers (Hogarth 1993b). Frobisher had exceeded by at least 127, the complement regulated by Council (*instructiones* in *SM* 2: 156). The voyage was first and foremost a mining venture, whereby at least 800 tons were to loaded, "of such oare as [they] already [had] from there this last yeare, or rather richer if [they] cane fynde the same". A secondary purpose was to plant a colony 100 strong to remain on Kodlunarn for the winter.

The complement included 156 miners (Table 2), almost all from the West Country, but three had been miners in the second voyage. Each was guaranteed 20s a month (estimated over 7 months), considerably more than a Cornish tin miner could expect for an equivalent period of time (cf Rowse 1969: 61). However the composition of these 'miners' was somewhat mixed; it included four blacksmiths, two special pursers, two miners-musicians, a miner-cook, a miner-minister, and a 'geometrician' (land surveyor, *EOD*), a roving draftsman who, no doubt, measured and prepared plans (now lost) of the main workings. One miner, Roger Heale, was awarded double pay (40s a month) and probably was mine captain: he would have supervised large gangs of miners (up to 100 at a single occurrence). A repeater from 1577, Richard Taylor, was also paid an extra high wage (26s 8d a month) and must have commanded a special position.[33] Also aboard were four assayers supervised by a master assayer, Robert Denham.[34] With them were the makings of at least three on-site furnaces (possibly eight) and three tons of miners tools, mainly for breakage. Some of the tools were heavy, e.g. 22 crowbars of 30 lb, 10 sledges of 20 lb, 35 wedges of 8 lb. Also provided were 920 baskets to carry ore, 40 one-gallon leather bags for drinking water, 20 wheel barrows, 20 hand barrows, 2 anvils of 2 cwt, 105 tons of coal and 7 pairs of bellows (Hogarth 1993a).

After the winter's preparations, the fleet weighed at Harwich (May 31), sailed through the Channel (June 1-4) and passed Cape Clear, southwest Ireland (June 7).[35] Then they skirted the southernmost tip of Greenland (June 19) and proceeded up the west coast. The morning following, "being under Friseland", Hall and Frobisher in one boat and Fenton in another rowed ashore. They examined the Greenlanders' tents, stole two dogs and noted that the rock held "the mother of rubie like to that in Meta incognita".[36] The fog descended and they "cast offward to the west" (June 21).

At last Greater Hall's Island came into view (June 27). The fleet passed from north to south across the North Foreland but the 'Straights' were blocked with ice and the fleet was forced to remain seaward. It was during a vain attempt to enter (July 2) that the *Bark Dennis* "chaunced to hyt an yse". She went down within the hour but all hands were saved. The remaining ships slowly drifted southward, in the mists, which parted on rare occasions, giving fleeting glimpses of the land. In effect, they were lost.

Then about midday, July 9, a well defined cape loomed on the horizon. In the words of the chief pilot, Christopher Hall:

> my Generall made yt to be the straicts and all the rest of the cumpany ... made
> yt the streicts, as the Generall sayd, and I stode against them all, and sayd yt was
> not yt... And he [Frobisher] was in a great rage & sware by Gods wounds that
> yt was yt, or els take his life. So I see him in such a rage, I toke my pinas &
> came abord the Thomas Allin againe.

In truth, they were below the Queens Foreland (as Hall suspected) and were viewing a channel separating Baffin Island from Ungava.

As result of the ensuing confusion, Frobisher led nine ships sixty leagues up Hudson Strait. By the time he had returned to Frobisher Bay, eighteen precious days had been wasted. In the meantime Fenton brought the *Judith* and *Michael* into the Countess Sound and discovered 'black ore' at 'Skipwith Mount' [Napoleon Bay], 'Fenton's Fortune' [near Sharko Peninsula] and 'Jones Island' [also near Sharko Peninsula].

The grand reunion of ships took place July 31: into Countess Sound sailed the *Beare Leicester*, *Francis* of Foy, *Salomon* of Weymouth, *Armonell* of Exmouth, *Emanuel* of Bridgwater, *Hopewell* and the *Ayde*. Two days later they were joined by the *Gabriel*.

The hundred - man colony

The hundred-man colony was another milestone of the Frobisher voyages. It was the first serious attempt by the English to establish a colony in the Americas.

The seeds of the concept were sown in 1577, for in the *instructions* for the second voyage[37], we find the following two items, which resemble closely the pertinent passages in the *instructiones* for the third:

> Item.   To consider what places maye be the most aptest to make fortification, yf neede requyre, to the defence of the moyeners [miners] and possessynge of the Contrie, and to bring perfect plattes and notes therof.

> Item.   Yf it bee possible, you shall leave some persons to wynter in the straight, givynge them instructions how they maye observe the nature of the ayre [air] and state of the Contrie, and what tyme of the yeare the straight is most free from yse [ice]. W[i]th whom you shall leave a sufficient p[ro]portion of vittals & weapons, & also a pynnesse w[i]th a carpenter and thinges necessarie, so well as maye bee.

Then, in Fenton's journal, appears a detailed list dated March 8, 1577 (1578 n.s.), where items are either identical or correspond closely to the vast majority of purchases tabulated in the Huntington manuscript (see Table 3).[38] This close correspondence tells us that plans for the colony were well advanced in the beginning of March 1578. The Fenton list also serves to isolate items that were earmarked for the 100 men, from those to be used by Frobisher's company. On March 11, Walsingham requested approval of Burghley and Sussex for a third voyage and setting up the aforementioned colony. However, at that time the Queen had already given her blessing to the scheme: "Hir ma[jes]tie [was] well pleased that a third viage be taken in hand.... and some provision be made for a 100 men to inhabit those north - west parts".[39] Official permission and advice to "procede w[i]th as convenient spede as ye may" was given March 12,[40] probably as a rubber stamp to what was already in common agreement. Frobisher received his patent to press ships and men on March 21[30] and, with Edward Harvey (his future lieutenant and captain of the *Gabriel*), hurried into the West Countrey to fill the roster of both Fenton's company and his own.[41] The ship's company was, for the most part, paid from April 1.

Table 3: *Mining tools and armaments for the 100 - man colony*

Fenton's journal (*PL* 2133, ff. 9-11)			Lok's Accounts (*HL*, HM 715, ff. 9-11)		
			Miner's tools		
baskets	400		baskets	800	
shovels	120		shovels	120	
spades	36		spades	36	
crowbars	30		crowbars of 30 lb	22	
mattocks	100		Mattocks	100	
pickaxes	30		pickaxes, 8-lb heads	30	
helves for pickaxes	300		*		
sledges of 24 lb	12		sledges of 20 lb	10	
sledges of 8 lb	12		sledges of 10 lb	10	
wedges	100		wedges of 0.6, 5 and 8 lb	101	
			Armaments		
corn-powder		2400 lb	corn powder		2400 lb
match		1000 lb	match		900 lb
rammers and heads	24		rammers and heads	24	
calivers	60		calivers	28	
longbows	60		longbows	60	
arrows for longbows	4320		arrows for longbows	4320	
crossbows	6		crossbows and arrows	**	
bowstrings	720		bowstrings	1440	
black bills	50		black bills	40	
pikes, long	50		pikes long and short	68	
targets	50		targets	60	
jacks	80		*		
lead		1000 lb	lead (sheet, sounding, fishing)		507 lb

* item not listed.
** quantity not specified; bought for £5 total.

Glossary (after *OED*). *Black bill*, weapon with concave blade, spike at back, spear head, and black handle; *caliver*, light musket or harquebus with ⅔ -inch bore; *corn-powder*, granulated gunpowder; *jack*, coat of mail; *match*, fuse for fire arms; *mattock*, digging implement, with a steel head, having on one side an adze and on the other a pick; *pickaxe*, single-tined pick; *rammer*, ramrod for a caliver; *target*, light round shield or buckler.

Frobisher's official *instructiones*[31] read:

> you shall Leave to remayne and to inhabite in the Lande nowe called *Meta incognita*, under the charge and government of Edward Fenton, gent, your Lieutenante gen[er]all, the Gabriell, the Michaell and the Judethe, w[i]th fortie hable marion[er]s, gonners, shipwrights and carpentars 30 soldiers and 30 Pyoners [miners], w[i]th sufficient vittale for xviij monthes for their provisione, releif, and maynten[an]ce.

Although not specified in the *instructiones* it is elsewhere clear that the miners were expected to excavate and stockpile further ore (in one estimate, 2000 tons in addition to ore mined in the summer of 1578)[42] and have it ready for pick-up by Frobisher's ships, who were scheduled to remove the colonists the following summer. The notion that the 'pyoners' were expected to break, lift and move tough, heavy rock from open pits during severe winter conditions, was indeed naïve and was probably the creation of the Queen, or the commissioners, or Lok, none of whom had acquaintance with hardrock mining or arctic conditions.

Frobisher, a man with experience in these northern climes, vigorously opposed the plan and "divers tymes with hevie countenaunce would cast out speches cullerablye, somtymes saying that this great furniture of buildings for C Fenton would be to littell purpose, somtymes that they shuld hardly be able to live there, and plainlie said to Charles Jackman at Harwiche [*c*. May 28] that they should not inhabit there".[43]   But these objections had little effect. Frobisher was not in supreme command of the expedition, but rather a servant of the Cathay Company, from which he received a regular wage. He had to answer to the Company, which, in turn, was under orders from the Privy Council, the official spokesman for the Queen. It is quite understandable that a rift developed between Edward Fenton, who strongly defended the concept of the colony, and Martin Frobisher, who opposed it with an equal force.

In the *instructiones*, Frobisher was requested:

> to search and consider of an apte place [to] best plante and fortifye these .C. men [left] behind to inhabite there as well against the dainger and force of the natyve people of th[aj]t Contrey and any other th[a]t shall seke to arryve.

The location of the fort was, therefore, not specified.

During the months following, Fenton was to be administrator, naturalist and meteorologist. Lok writes: [31]

> You shall leave w[i]th capten fenton yo[u]r Lieuten[a]nte generall the government of those .100. p[er]sones to remayne in that Countrie w[i]th instructions howe he maye best observed the nature of the ayre and may discover & know the state of

the countrie *fro[m] tyme to tyme* [Lord Burghley's hand], and what tyme of the yeare the straight is most free from eysse [ice], *Kepy[n]g to th[at] end a journall wikly of all accts* [Lord Burghley's hand], w[i]th whome you shalle leve the Gabriell, the Michaell and the Judith w[i]th suche p[ro]portion of victualls and other necessarie things as are alredye appoynted to him and his companye for that purpose, supporting his want w[i]th able & skyllfull men for that purpose & w[i]th other things necessarie w[hi]ch you or any other of the shippes maye conveniently spare at yo[u]r reatorne.

Mining, which in the intial plans was given a major role, had dwindled by the time of the third voyage. Thus we find only 5 miners and 6 labourers listed in the Huntington manuscript[44], a drastic cut from the '30 pyoners' in the *instructiones*.[45] Mining tools (Table 3) were implements of breakage and were modelled after those used in the second expedition; they were mainly acquired from the Tower and Richard Lane, a blacksmith. The 'pyoners' had no mine captain and no assayers. Obviously they intended to break and pile rock on a small scale, as their sole mining endeavour.

Why then was this colony planned? An answer can be found in a statement of Lok in the instructions:

You shall well consider what place may be most aptest further to fortify upon hereafter (yfe nede requier) bothe for defence of the myners and also for possessinge of the countrie...

The last reasons, those of establishing a claim to possession and protection of the mine sites, may be regarded as the true objectives. The colony, in its original form, was to be a one-year venture and mining, a secondary pursuit.

The structure, itself, probably represented the first attempt to set up a prefabricated building in the high north. It was described as a "strong forte or house of timber, artificially framed, & cunningly divised by a notable learned man here at home, in ships to be carried thither" (G. Best in *SM* 1: 81). Elsewhere, the compound was described as "a howse of timb[e]r hewen framed for o[u]r lodging & storehowse conteyning 132 foote in length and 72 foote in bredthe, with ij [Longehowses] at either ende thereof [and] wth iij other bulworks of defence thereunto adjoyninge".[46] Cranes and pegs ['pinnes'] were provided for hoisting. The timber was English elm, the roof was to be tiled, the walls bricked and mortared, the stoves fueled with coal and wood.[47] The carpenter in England (probably the "learned man at home") was Thomas Townson; his bill for timber, design and construction, alone, totalled £ 246 17s 6d.[48] Surprisingly, no provision seems to have been made for armament, other than weapons of hand-to-hand combat (Table 3), although light cannon on the ships (sacres, minions and faucons) might have been transferred to the fort for this purpose. Come winter, the *Judith*, *Gabriel* and

Michael were to be hauled ashore by means of specially constructed claws ('crabes') attached to capstans.[49]

At departure from England, the ships were loaded with equipment for the proposed colony. The purser, Nicholas Chancelor, was in charge of distribution. Unfortunately, half the framed house went into the *Dennis*, which sank off Resolution Island, July 2, and half the bricks (10,000 of them), much of the lime (at least 320 barrels), some 13 cartloads of elm roofing, and a quarter of the firewood (in total 5000 billets) went into the *Thomas*, which deserted about August 8, without reaching the Countess Sound. In addition the four "crabes w[i]th their furniture" and most of the longbows, bills and pikes were in the *Mone*, which arrived at the north shore on the eve of departure and never anchored in Countess Sound. But, worst of all, was the loss of 84 tons of beer, a staple of the time, gone with these three ships and the *Anne Francis* (another late arrival to the north shore). A crisis now arose. What was to be done? On August 9, Frobisher convened his chiefs of staff (his 'councell') on Kodlunarn. The south and east sides of the house were intact and available, but much of the remainder was gone. Fenton offered to overwinter with a smaller group (50) in a reduced dwelling, but time and materials were against him, "wherefore, it was fully agreed uppon, & resolved by the General and his counsell, that no habitation shoulde be there [that] yeare" (George Best in *SM* 1: 105-6).[50] Thus ended the ambitious attempt of the English to set up a colony in the Arctic.

Baffin Island 1578: mines and loading

Frobisher now established basecamp on 'Countess of Warwick' [Kodlunarn] Island (Fig. 8). On July 4 Frobisher published and proclaimed his "orders for the government of the whole fleete", which were formally read aloud after a blast on the trumpet. Mining was quickly resumed, first at the Reservoir Trench on Kodlunarn (the Ship's Trench having been finished in the second voyage). Then, when the rock became too tough to break, the mine was abandoned and the miners were transferred to Fenton's Fortune, on the east side of Countess of Warwick Sound.

The "riche redde ewre" of Jonas Mount, sampled as sand in the second voyage, was hunted in vain on July 30, August 4 and August 27. The failure of discovery must have been a great disappointment to Frobisher because, apparently, he believed that there were considerable deposits available and, after the very high assays made in November and December 1577,[19] the company had pinned its hopes on this commodity. Frobisher had, therefore, included Gregory Bona, who "knewe the place", in the complement of the third voyage.[51]

August 10, Countess of Sussex Mine, the largest and reputedly one of the richest deposits, was discovered on the main, six miles west northwest of Kodlunarn. The ore (black, red and mixed) was in good supply and easy to extract, all to the good, as the

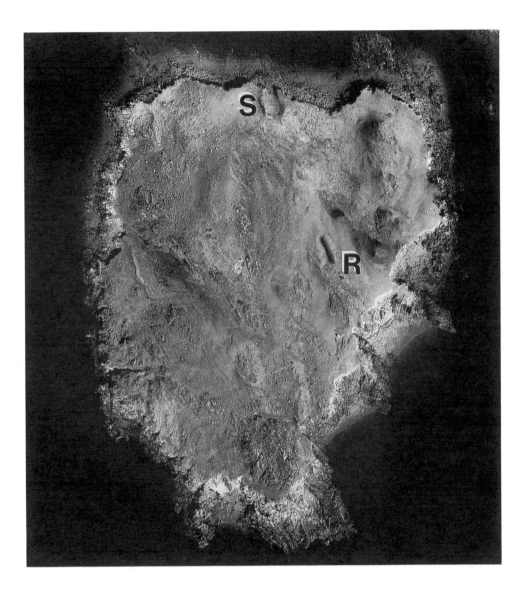

Figure 8: Aerial view of Countess of Warwick Mine (Kodlunarn Island), Frobisher's mine of 1577 and base of operations 1578, showing the Reservoir Trench R, and Ship's Trench S. The long diameter of the island is approximately 500m. Photograph A27161-170 of the Department of Energy, Mines and Resources of Canada, taken in 1987. Published with permission.

mines were surface trenches put down by brute force and without the use of blasting powder. Denham was pleased with the assays.[52] This became the centre of mining for the summer. It was also the locale of a surprise appearance; Captains Upcot and Best, rowing a boat precariously held together with a few nails, suddenly appeared at the mine on August 22. They had crossed the 'Straights' and brought good news that the *Anne Francis* and *Mone* were safe and sound on the south shore.[53] The mine was finished August 27, but Fenton and company returned August 31 to collect miners' tools and one of the *Ayde's* sails. They also buried various supplies in anticipation of a future visit (E. Fenton in Kenyon 1981). Denham's Mount, another important mine, was discovered August 16 in 'Dyer's Sound' [Victoria Bay?]. It held four types of 'black ore'.

The last mines to be worked on the north shore were those of Beare Sound, where sporadic loading took place throughout August and into September. It seems that several mines were worked, all in the vicinity of the entrance to Beare Sound, but the exact locations are not known; certainly all mines were exposed to the sea and provided difficult anchorages. The first operation began August 2, when Frobisher, with two pinnaces, set out to fetch ore. In all likelihood, this was from 'Leicester's Island', a deposit discovered and stockpiled the previous year. This must have impressed the General, for on August 9 he set out with Christopher Hall, the *Gabriel*, the *Michael* and at least four pinnaces and a hundred miners. The tiny barks were used to relay ore to the cumbersome *Ayde*, which lay safely at anchor in the Countess Sound. Then, when the traffic was clear, the *Thomas Allen*, *Salomon* and *Emanuel* came in for loading.

In the meantime, the *Anne Francis* and *Mone* loaded at 'Best's Blessing', a dark rock off Resolution Island. Here they found "blacke Ore of the same sorte whiche was broughte into Englande thys last yeare ... that might reasonably suffice all the golde gluttons of the worlde" (George Best in *SM* 1: 111). Best brought several types of this rock to the north shore, where on August 22, "Denham drew an assay or twoo". The results were disappointing (E. Fenton in Kenyon 1981: 198) and, on the following day, four additional proofs were only "reasonably well liked of Denham" (E. Sellman in *SM* 2: 67). As the ships were now nearly loaded, the matter was let pass. Finally, the laden ships crossed to the northland where, on August 30, the *Anne Francis*, with eight great leaks, was hauled ashore and repaired.

Mining in 1577 and 1578

Frobisher's mines were shallow open pits, labouriously excavated in exceedingly hard rock (biotite gneiss and 'black ores') at great effort, without the benefit of explosives. Amongst the "Implemets for mynes of New Land", Lok itemized flat *plates* and *wedges*.[54] These show that the ancient 'plug and feather' method of wedging was employed on Baffin and Kodlunarn, (see Chapt. 1). *Chisels, picks, sledges* and *crowbars*

Table 4: *Tonnage of ore shipped in the third voyage (1578)*

	Ship	Captain	Tons burden	Localities and weights (tons) Established by Sellman							Official Total
				1	2	3	4	5	6	7	
Company's ships	*Ayde*	M. Frobisher	200	20	110	-	-	-	-	-	132
	Judith	E. Fenton	100	80	-	-	-	-	-	-	60
	Michael	M. Kindersley	25	-	20	-	-	-	-	-	13
	Gabriel	E. Harvey	20	-	20	-	-	-	-	-	9
	Thomas Allen	G. Yorke	160	100	60	-	-	-	-	-	166.3
	Hopewell	H. Carew	150	-	-	140	-	-	-	-	136.2
Commissioned ships	*Anne Francis*	G. Best	130	-	-	-	130	-	-	-	116.0
	*Thomas**	W. Tanfield	130	-	-	-	-	-	-	-	-
	Francis	H. Moyles	130	80	-	-	-	50	-	-	126.8
	Mone	- Upcot	100	-	-	-	100	-	-	-	98.2
	Beare Leicester	R. Phillpot	100	-	-	100	-	-	-	-	84.0
Non-commissioned ships	*Salomon*	H. Randall	120	60	60	-	-	10	-	-	100.4
	Emanuel+	R. Newton	100	30	60	20	-	-	-	-	-
	Armonell	T. Courtenay	100	85	-	-	-	5	5	5	94.2
	Bark Dennis++	A. Kendall	100	-	-	-	-	-	-	-	-
			455	330	260	230	65	5	5		1136

Localities: 1. Countess of Sussex Mine, 2. Beare Sound, 3. Denham's Mount, 4. Queen's Foreland, 5. Countess of Warwick Mine, 6. Winter's Furnace, 7. Fenton's Fortune.

Estimated tonnages from E. Sellman in *SM 2*: 69-70; official weights from *PRO*, E 164/35, f 211; E164/36, ff. 67-85.

* Deserted, August 8
+ Discharged ore in Ireland
+ + Sank near Resolution Island, July 2.

were also used. The well developed foliation of the wallrock (gneiss) surrounding thin but massive payzones of 'black ore' would have enhanced the splitting operation.

It is surprising that, after enduring four centuries of natural erosion, enhanced by extreme weather fluctuations, there still remains clear evidence of breakage on rock surfaces in the Baffin Island area. Kenyon (1975a: 45; 1975b: 143) illustrates plucking and chipping of the wall rock at one of the trenches on Kodlunarn Island.

> The chipping technique was generally used to fetch material off the surface, i.e.
> vein stuff, and the *long-handled picks* in this operation had a wide range of
> double-pointed blade sizes, weighing from ounces to several pounds; they were
> capable of removing rock with great delicacy to avoid dilution of the ore. The
> shorter handled *pickaxes* (used for dressing rock) had a pick at one end of the
> head and a crude cutting axe at the other. *Mattocks* (used in chopping rather
> than the hacking of the pickaxe) may have been pick-like but with a wider chisel
> blade than the 'axe'. Pickaxes and mattocks would have been useful tools in the
> trenches.[55]

Distinct from these markings (chippings) are narrow grooves on the footwall gneiss bordering the payzone of the Ship's Trench, Kodlunarn Island, which are not strictly linear but have been deflected at irregularities in the rock (Fig. 9). They are therefore not borings but represent grooves formed by the miner's *single-tined picks*. In fact, the grooves are:

> characteristic of *hammer-gads* or *short-handled picks* with a hammer head and
> point, much like an enlarged geological hammer of today. The point was
> inserted into a crevice in the rock and hammered down to form the groove,
> crudely splitting off the rock in front of it.[55]

The operation is illustrated in Figure 4. *Bergeisen* and *ritzeisen* picks (Chapt. 1) must have been used by Frobisher's miners. It is surprising that the discarded heads have not turned up amongst the archaeological relics.

Shattering due to explosives was not evident at the Frobisher mines, but this is hardly surprising, because in the sixteenth century blasting powder was not employed by Europeans. Gunpowder (including cornpowder and serpentine), listed in the business accounts of all three Frobisher voyages, was therefore destined for munitions. A small purchase of saltpetre by Robert Denham was not provided as a constituent of blasting powder but rather an ingredient of nitric acid, [56] known to be employed by Denham in Baffin Island, to part silver from gold, or possibly as an oxidant in the additive to the crucible charge (Hoover & Hoover 1950: 237-8, 439n). Likewise, no evidence could be found for breakage by fire-setting and quenching (Chapt. 1), a method that produces a distinctive shatter pattern in brittle rocks.

Figure 9: Pick scars on footwall gneiss, Ship's Trench, Kodlunarn Island.

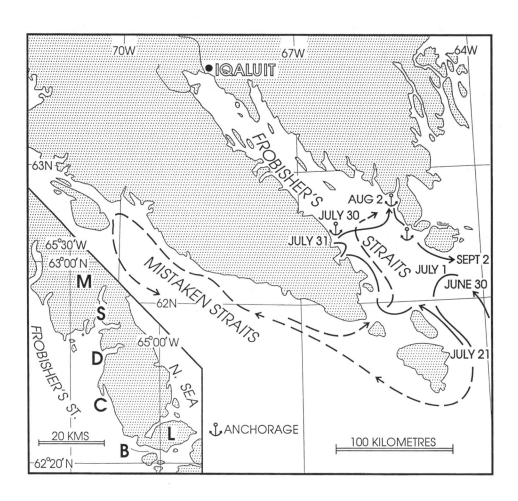

Figure 10: Frobisher's third voyage (1578), showing routes of the *Thomas Allen* (coarse broken line); *Ayde*, *Beare*, *Francis*, *Hopewell*, *Emanuel*, *Salomon* and *Armonell* (fine broken line). Inset shows Frobisher mines (crossed picks). Abbreviations: B = Beare's Sound, C = Corbett's Point, D = Diare's Sound, L = Loks Land, M = Meta Incognita, S = Seate Gulph (all Elizabethan names).

Return voyage, 1578

Departure from Beare Sound took place at the beginning of September (Fig 10). The ships made their way to England and proceeded through the Channel with no difficulty (excepting the *Emanuel*, as discussed in Chapter 3). At Portsmouth, Frobisher left the *Gabriel* and proceeded to London by land, leaving supervision of landing in the hands of Fenton, now aboard the *Ayde*. On October 12, Fenton's company apprehended one Thomas Halpennie "in a little Barke of 50 toones taken from a Britaine". This bark was manned by Fenton's mariners but, in spite of this precaution, Halpennie with three others escaped, having been set on land at Tilbury by their own boatswain. The incident caused a three-day delay in landing. The *Ayde* cast anchor off Dartford, October 17.[57]

Frobisher had made good his resolve to bring to England at least 800 tons of ore: the official register is given in Table 4 (1136 tons) including 195 tons from the non-commissioned ships *Salomon* and *Armonell*, tonnage the Company considered their own. One hundred and ten tons remained in Ireland from the wrecked *Emanuel* (see Chapter 3).[58]

It seems that all the loaded ships received some compensation for ore, but full payment was made in the cases of the *Beare* and *Salomon* only. Lok even tried to arrange payment for 16 tons of 'dead freight' in the *Beare* but had to settle for 4 extra tons.[59] Final payment for the *Salomon* was made in the fall of 1580, but by that time the Cathay Company was penniless and money was siphoned from another source.[60]

We have no details of the on-site assays, although in Fenton's journal alone there is reference to assays made on 18 days between July 31 and August 22. The *booke of register*, reporting all assays in accordance with a clause in the instructions, never reached England. According to Frobisher it went "into a benche in his Cabben of the Shippe the *Ayde* and ... in a storme of wether [the] Cabben was brast open and so all is gonne". The incident occurred (in the absence of Frobisher) in the early hours of September 15 and was corroborated by Christopher Hall:

> The winde .... came to the West SW & a great storme. The ship sponed [spun] afore the sea, without any saile, NNE & at tht present 4 of [the] clock, the sea be[a]t in at my Gen[er]alls cabban, & burst from the Cabban floors to the Windows al the Timber & burds [boards] unto him who was at the helme; his name [was] Fraunces Austen.[61]

Frobisher was to bring "a Dooble of this book to be made and brought home in an other shipp",[62] a request he ignored and the commissioners forgot.

Denham claimed the ore was high grade: "for the most part of the ships lodinge ....
holdithe almost an onse of gold in C of ewer" [one ounce per cwt], but Lok was
skeptical: "I do not believe ytt untill I see better proffe".[63] With data at hand, it became
inceasingly apparent that the ore had little value.

Back in England, there was much skepticism on the worth of the ore. Frobisher's task
now was to re-establish its value and, therefore, his own credibility. The voyages had
not come cheaply; they had cost £19,200, two ships, about 22 boats and pinnaces, and
at least 24 lives.[64] It was necessary to show convincingly that gold was present in
important quantities. High- grade assays were mandatory.

Notes

1. Frobisher seems to have adapted easily to the transition from Passage to prospecting. He
appears to have made no attempt to unload his holdings in the venture (£400 including the final
assessment) and redirected his attention to mines and metallurgy with enthusiasm.

2. The ships' company in all three voyages has been listed by name and, for the most part, by
ship assignment and duty in a table compiled mainly from paylists (Hogarth 1993b).

3. Unless otherwise noted, the information of this voyage has been compiled from the accounts
of George Best (*SM* **1**: 46-51), Christopher Hall (*SM* **1**: 149-54) and Michael Lok (*SM* **1**:
157-66).

4. The number of Michael Lok (*BL*, Cotton MS, Otho E/8/8, f. 43 V 1577; *SM* **2**: 82).
Unfortunately the list of personnel referred to by Lok is no longer with the manuscript. A list,
mainly from Exchequer Book *PRO*, E 164/35, with names, assigned ships and duties, giving 36
officers and men, has been compiled by Hogarth (1993b).

5. *The Doynges of Captayne Furbusher, BL*, Landsdowne MS 100/1, f. 2R.

6. The grid reference to Little Hall's Island (as now placed on Canadian topographic map
25-I/15-L) is, in NTS coordinates, 20 V ME 4032. This island is, however, 1.5 mi long and 0.6
mi wide max. A more plausible location is a smaller island at 20 V ME 3533.

7. A Robert Garret (also spelled Garrat) was on the payroll for the first voyage, but he was paid
on the return (*PRO*, E 164/35, f. 27) as well as the ingoing voyage (ibid. f. 18). This, however,
may be an instance of fraud.

8. M. Lok to F. Walsingham. *PRO*, SP 12/122/25, April 22 1577; *SM* **2**: 84.

9. The syllabus of the 'Company of Kathaye' is laid out in *PRO*, SP 12/110/21, 22, 1577; *SM* 2: 103-7. As pointed out by Manhart (1924: 46-7) and Shammas (1975: 96), there is no evidence that the company was ever incorporated.

10. The debit is recorded in *PRO*, E164/35, f. 127. "paid to m[aste]r George Winter for rent of the storhouse, for 28 wekes at 5 sh [the weke] from the 15 [of] october to [the] laste of Aprille 1577 ... li 7.0.0".

11. Unless otherwise noted, the information on this voyage has been compiled from the account of George Best (*SM* 1: 52-79).

12. These *instructions* were first published by Ellis (1816) and were reprinted in *BL*, Additional MS 39852. They first appeared as manuscript in *PRO*, SP 12/113/13, May 22 1577, were copied and revised as *PRO*, SP12/113/12 [May 1577], and preserved as a one-page abstract in *PRO*, SP12/113/14, May 22 1577. They reappeared in the 17th century as *NMM*, Croker MS/Rec. 3, ff. 55-9. They were printed by Collinson (1867: 117-20) but were omitted from Stefansson & McCaskill (1938).

13. The burden quoted by Anderson (1959: 12), *viz.* 255-319 tons, appears large when compared with contemporary data (150-250 tons). According to Anderson the ship was completed in 1562, measured 73 feet on the keel and 25 feet on the beam, and fully armed carried 30 guns. She was condemned in 1599. She was probably rebuilt and enlarged after the siege of Smerwick (7-10, November 1580).

14. The number that finally sailed, as quoted in the Exchequer papers, was 143; *BL*, Lansdowne MS 100/1, f. 3V, gave 145; names gathered by Hogarth (1993b) totalled 146. Of these numbers, 8 were miners (7 from the Forest of Dean, 1 from Cornwall), 3 were 'goldfiners' [assayers] and 3 were 'smithes' [blacksmiths]. The miners, assayers and blacksmiths were supervised by Jonas Shutz, himself an assayer.

15. According to George Best (*SM* 1: 57) the promintory forms the eastern extremity of Greater Hall's Island. This island was said to be somewhat smaller than the Isle of Wight, and was divided from the "continente of the norther land by a little sounde called Halles Sounde". These data fit Loks Land better than the geographic position in which Hall's Island is now placed.

16. *The Doynges of Captayne Furbusher, BL*, Lansdowne MS 100/1, f. 2V.

17. Of the ore on Kodlunarn Island, Best reported that "there was store and indifferent good". Here they loaded rather "than to seeke further for better, and to spend time in jeoperdie" (*SM* 1: 64). Best's statements suggest the ore was not regarded as high grade but was loaded regardless, rather than return empty handed. The same doubts are echoed by Dionyse Settle, who was also present on the voyage: "There is much to be said of the commodities of these

Countries, which are couched within the bowels of the earth, which I let passe till more perfect triall be made there of" (*SM* 2: 24).

18. Located from information given by Edward Fenton (Kenyon 1981: 188, 199).

19. B. Kranich to F. Walsingham *PRO*, SP 12/118/43, Nov. 26, 1577; *SM* 2: 139; W. Winter and M. Lok to F. Walsingham *PRO*, SP12/119/8, 9, Dec. 6, 1577; See also *The Abuses of Captayne Furbusher*, *PRO*, SP 12/130/17, Mar. 25, 1579; *SM* 2: 208, and *The Doynges of Captayne Furbusher*, *BL*, Lansdowne MS100/1, f. 6R. The red and yellow ore described as a sand in the *Doynges* was probably gossan. The solid red, and locally common granite pegmatite and feldspathic quartzite are unlikely identities of 'red ore' (Kenyon 1975a: 40, 1975b: 158). No sample of the original 'red ore' is known today.

20. Official tonnages of ore weighed in England are from *PRO*, E 164/36, f. 327, 1581; cf. 200 tons estimated in Baffin by Best (*SM* 1: 75) and Dionyse Settle (*SM* 2: 18) and 124 tons in the *Ayde* and 16 tons in the *Gabriel* estimated in Bristol by Fenton (*PRO*, SP 12/118/40, Nov. 25, 1577; *SM* 2 134).

21. *APC* 10: 38-40, Sept. 18, 1577.

22. Lok's memorandum, *PRO*, SP 12/115/35, Sept. 1577; *SM* 2: 109- 10. This was followed by Council directives: *APC* 10: 54-6, Oct. 17, 1577; *ibid.*: 273, July 3 1578. Additional details on the transport and storage of ore are given by Michael Lok in *PRO*, E 164/35, f. 149 and E 164/36, f. 327.

23. *APC* 10: 44, Oct. 2, 1577. Winter must have been efficient in this operation for a thorough search has revealed neither manuscript nor book in London and Spain.

24. For further information on this illustration see Hulton (1961, 1984). It was inspired by an unknown artist and redrawn by John White, Governor of the second Virginia Colony of Sir Walter Raleigh. It is unlikely that the original sketches were made by White himself, who is not recorded in the paylists (*PRO*, E 164/35), which appear to be complete. The fate of the Inuit is considered in detail by Cheshire *et al.* (1980).

25. Information from M. Lok in *PRO*, E 164/35, 1578.

26. Estimate of M. Lok in *BL*, Lansdowne MS 100/1, ff. 3R-3V.

27. M. Lok, *BL*, Cotton MS/Otho E 8/8, f. 45R, Nov. 16, 1581.

28. *The Doynges of Captayne Furbusher*, *BL*, Lansdowne MS 100/1, f. 7R. The fact that this letter was never produced suggests that the allegation may have been fabricated either by Frobisher or the author of the document [M. Lok?].

29. *Advertycements owt of Russia, PRO*, SP 91/1/1a, Jan. 26, 1579.

30. *Calendar of Patent Rolls, Elizabeth 1*, 7(2886) Mar. 21, 1578.

31. Commissioners to Martin Frobisher, 1578, *Instructiones geven to o[u]r lovinge frind Martine Frobiser, Esquier, PRO*, SP 15/25/81; *SM* 2: 155-61. Text by Michael Lok, with marginalia, additions, and corrections penned by Lord Burghley. This is a draft; the final version has not been located.

At the time of drafting the *instructiones*, the *Beare Leicester* was apparently under consideration as a commissioned ship. The text begins:

> First, you shall enter as capten Generall into the charge and government of theis
> shippes and vessells, viz the Ayde, the Gabriell, Michaell, Judethe, the Thomas
> Alline, Anne Frauncs, the hoppewell, the mone, the Frauncs of Foy, the thomas,
> & the [blank].

32. The charter parties were made out in the name of Michael Lok. (Sussex, Leicester, Walsingham, Willson, Warwick and Knollis to the Commissioners of the Cathay Company, May 6, 1578, *PRO*, SP 12/149/42 (xi)). They were drafted and ratified by the Commissioners on May 16, 1578. (*PRO*, SP 12/149/42 (xii)) and delivered to the ships at Harwich.

33. Information on miners of the third voyage is scattered throughout the account books of Michael Lok and Edward Sellman but is especially detailed in the account book of Sellman (*PRO*, E164/36, ff.147-64). Three blacksmiths and three miners were unplaced in ships but were grouped together with John Page (a miner assigned to the *Gabriel* in the Huntington manuscript; *HL*, HM 715, f.16). All seven were with Fenton's company, having volunteered to overwinter in Baffin; they are included with the complement of the *Gabriel* in Table 2.

34. Robert Denham belonged to a well respected family of London goldsmiths and had been assayer on the second Frobisher voyage. He was paid £3 per month in addition to a pension of £50 per annum. Gregory Bona, a German assayer, had also been present in the second voyage and was the only person on the third voyage who had visited the site of the rich red and yellow ores. He was paid £14 5s in advance, but his name does not appear on later paylists. William Humphrey, the younger, was son of a former assaymaster at the Mint. He was paid 50s per month. Junior assayers, probably assistants, were John Lambell, paid 26s 8d per month, and Robert Peacock, paid 20s per month. All wages were guaranteed for 7 months (*HL*, HM 715, ff.19, 20; *PRO*, E 164/35, ff.185, 256; *PRO*, E164/36, ff.56, 57, 152).

35. Information for the voyage has been largely taken from the Journals of Christopher Hall (*BL*, Harley MS **167**, ff. 183-200, 1578) and Edward Fenton (*PL* 2133, ff. 5-73, 1578; in part published by Kenyon 1981), and the account of Edward Sellman (*BL*, Harley **MS** 167, ff. 165-180, 1578; *SM* 2: 55-73).

36. Probably garnet. On the second voyage, Lok (*BL*, Lansdown MS 100/1, f. 6V) reported that Jonas Shutz had discovered "a great Rubie stone of more than one ynche square, a present mete for a prynce". This was almost certainly almandine garnet.

37. The quotation is taken from items 10 and 11 in the *instructions* for the second voyage (*PRO*, SP 12/113/12). Almost identical are items k and m in the draft (*PRO*, SP 12/113/13) and lines 18 to 23 in the abstract (*PRO*, SP12/113/14.)

38. The two manuscripts referred to are the Journal of Edward Fenton, 1578 (*PL* 2133), and *Thaccount of Michael Lok*, August 31, 1578 (*HL*, HM 715). In Fenton's (preliminary?) list some items, like baskets and bowstrings, were grossly underestimated and were simply doubled in Lok's orders, whereas others, such as crowbars and calivers, were less notably in short supply, and the deficit was probably made up by materials remaining from the second expedition. However, a £120 debit of Frobisher for "calivers furnished, targatts, bowes and arrowes, and other provicons of armor for 100 men to inhabit the new land" was disallowed by the Commissioners because, at the time of audit, (1581) "all armour remayneth still in Captain Furbushers handes". *PRO*, E164/35, ff. 297-305. These 'calllivers' do not appear in Table 3.

39. F. Walsingham to the Lord Treasurer and Lord Chamerlayne, March 11, 1578, *PRO*, SP 12/123/7; *SM* **2**: 131.

40. Council to Sir W. Wynter, Martin Frobisher and others, March 12, 1578, *PRO*, SP 12/149/42 (xiv).

41. *HL*, HM 715, f. 23; *PRO*, E 164/35, ff. 287, 288, 290, 297, 299.

42. Anonymous, May 3, 1578. *PRO*, SP/12/124/1; *SM* **2**: 152-3.

43. Quotation [M. Lok to Commissioners, 1581 ?]. *BL*, Lansdowne MS 100/1, ff 7V-8R.

44. *Thaccount of Michael Lok*, August 31, 1578, *HL*, HM 715, ff. 15-21. This section of the manuscript has been transcribed by Parks (1935: 187-8) and repeated by Stefansson & McCaskill (*SM* **2**: 220-1).

45. Another grouping is given by Edward Sellman in his account: *PRO*, E 164/36 f. 98, *viz.* mariners 32, fishermen 6, ship carpenters 2, miners 28, ministers 1, bakers 2, tailors 2, shoemakers 2, coopers 2, carpenters 2, bricklayers 2, surgeons 2, soldiers 15.

46. The quotation is from Fenton's journal: *PL* 2133, f.12, with information in parentheses derived from *HL*, HM 715, f. 12.

47. Most of this information was derived from *PL* 2133, ff. 9-12, and *HL*, HM 715, ff. 9-11. In the outgoing ships were 20,000 bricks, 2000 tiles, 2000 [barrels ?] of unslaked lime, 80 chaldrons [120 Tons] of coal, and 2000 stove-sized billets of wood. Also, there were 2 large stoves bought from Jeronimus, a miner, and 8 small stoves bought from Jacob Johnson, a blacksmith, all probably destined for the fort (cf. 4 stoves in Fenton's journal).

48. *Thaccount of Michael Lok*, August 31, 1578, *HL*, HM 715, f. 12V and *Thaccountt of Michael Lok*, December 31, 1578, *PRO*, E164/35, f. 179. At the end of 1578, £46 17s 6d were still owed to Townson.

49. *Thaccount of Michael Lok*, August 31, 1578, *HL*, HM 715, ff. 8R, 11; *Accoumpt* [of E. Sellman], March 1579, *PRO*, E 164/36, f. 80. The *Mone* was held responsible for the loss of the 'crabes'.

50. Published sources used in this paragraph include G. Best in *SM* 1: 81, 102, 105-6; T. Ellis in *SM* 2: 43-4; E. Fenton in Kenyon (1981: 192). Considerable data have been derived from the audited accounts of the Cathay Company, especially those of M. Lok in *PRO*, E 164/35, ff. 183, 184; M. Lok in *HL*, HM 715, f. 12; E. Sellman in *PRO*, E 164/36, ff. 68, 70, 78.

51. *BL*, Lansdowne MS 100/1, f. 6R.

52. The principal assay site was on Kodlunarn Island, where there may have been more than one furnace. Another assay site was on the shore of Victoria Bay, a third at Winter's Furnace and, by late August, at Beare Sound. No assay furnace was erected at Countess of Sussex Mine.

53. The story is related graphically by G. Best (*SM* 1: 112-5), who was present on the *Anne Francis*.

54. The passages concerned are in M. Lok's account, August 31, 1578, *HL*, HM 715, ff. 10R, 11R: "paid to Rycharde Lane for 32 small plate wedgys [weighing in all] 20 lb" and "payd to Robert Crokey for 24 plates" [part of a shipment totalling 5 cwt 8 lb]. Lane and Crokey were blacksmiths who supplied mining equipment for the Frobisher voyages. The name 'feather' is derived from the German name for an iron plate, 'feder'. Wedges are also itemized in these two purchases, individualy weighing 5,7 and 8 lbs.

55. The quotations are taken from correspondence of Mr. L. Willies, Peak District Mining Museum (Matlock Bath, England), dated Dec. 12, 1991.

56. The item is listed in *HL*, HM 715, f. 9 "paid to officers in the Tower... [for] saltepeter j^c [1 cwt] to Robert denham". An item for the second voyage (*PRO*, E 164/35, f. 104) reads "paid [to officers in] the touer... for Salte peter xxviij lb [28 lb]" and may have been ordered for the same purpose.

57. The statement of Best (*SM* 2: 122) that "all the fleete arrived safely in Englande aboute the firste of October" is not correct.

58. Another total (not broken down according to ship) is given in *PRO*, E164/36, ff 327-9 as 951.8 tons in the commissioned and company ships, 194.6 tons in the *Armonell* and *Salomon*.

59. Final settlement to Richard Fairweather, the elder, joint owner of the *Beare Leicester* was £450, roughly equivalent to 88 tons at £5 2s 8d the ton (*PRO*, E 164/35, f. 280, Oct. 4 1580). Money came from the second and third stock levies of the Cathay Company. Lok had originally asked for £513 6s 8d (*PRO*, E 164/36, ff. 77, 1578). Richard Fairweather also requested, but was refused, £96 for provisions for 20 who had manned the vessel during the voyage (*HL* 715, f. 24, 1578).

60. Hugh Randall, owner of the *Salomon* of Weymouth was paid £60 19s 8d out of the third levy (*PRO*, E 164/35, f. 280, Oct. 4 1580) and £630 from fines collected for unlawful exports from Dorset (*Calendar of Patent Rolls, Elizabeth 1*, 8 (2162), Oct. 22 1580). This is £ 175 12s above the allotment for 100.3 tons of ore (equivalent to £60 19s 8d) and probably represents a consideration for unpaid, pressed mariners and/or ship damage sustained in the voyage.

61. Quotation from *The Doynges*, *BL* Lansdowne MS 100/1, f. 12R, and Christopher Hall. *The accownt of the third Voyage to Meta incognita made by Christopher Hall*. *PRO*, Harley 167/42, f. 197 V.

62. *Instructiones*, *PRO*, SP15/25/81; *SM2*: 159-60; Lord Burghley's interjection in margin of manuscript written by Michael Lok.

63. M. Lok to Lord Burghley, *CP*, 161/71, Oct. 10, 1578.

64. Expenses were derived from *Summary Accounts* (*PRO*, E164/36: 317-41) and exclude the cost of assays and the Dartford works; ships were the *Dennis*, lost July 2, 1578 (*BL*, Harley 167/42: f 190R), and *Emanuel*, lost early November, 1578 (T. Wiars in *SM 2*: 253 and G. Desmond *PRO*, SP 63/65/15); boats and pinnaces were lost July 1, 1575 (G. Best in *SM 1*: 47), 20 August 1576 (C. Hall in *SM 1*: 153), and September 1-3, 1578 (G. Best in *SM 1*: 121); 9 men were lost in the first voyage (G. Best in *SM 1*: 47,49), 2 in the second voyage (G. Best in *SM 1*: 79), and 13 in the third voyage (E. Sellman in *SM 2*: 65,66,68,72,73). No further deaths are recorded in the manuscripts of Fenton and Hall but the number of deaths does not agree with the statement of Best (*SM 1*: 122) that "there dyed in the whole Fleet in all this [third] voyage not above fortie persons". In a separate report, dated June 16, 1581 (*The causes and foundacyon of the iij voyages*, *PRO* SP 12/149/42 (iii), Lok apportions expenses as follows: first voyage - £1600, second voyage - more than £4,000, third voyage - £15,000.

CHAPTER 3

The *Emanuel* of Bridgwater and Ireland

The history of the *Emanuel* is a microcosm of the Frobisher enterprise and certain other events of the day; the ore, the ship and Smerwick Harbour (the graveyard of the *Emanuel*) are closely intertwined and cannot be easily separated. The lading was from Frobisher's three largest mines, but two of these, Beare Sound and Denham's Mount, cannot be located today. The ultimate fate of the ore is tied in with the history of this part of Ireland; in fact the very name of the site, Fort Dún-an-Óir or Fuerte del Oro (the fort of gold), derives from Frobisher's cargo (Hitchcock 1854), but with the advance of years, the association with the *Emanuel* is now forgotten. Dún-an-Óir is the locale which attracted various principals of the Frobisher enterprise: Fenton - the lieutenant, Carew and Courtenay-the captains, Winter and Pelham - the commissioners. They failed to note the presence of the *Emanuel* in the shallows; there were more pressing matters at hand. Now it seems appropriate to recall the Irish connection.

The *Emanuel* was a broad-beamed fishing vessel (a buss) with a burden of somewhat over 100 tons. She appears to have been cumbersome, lacking the manoeuverability of the accompanying barks. The appearance of this ship is not known but it probably bore resemblance to the smaller buss illustrated in Figure 11.

In the third voyage, the *Emanuel* was owned and captained by Richard Newton, a respected seaman from Bridgwater, Somerset. Newton, formerly a resident of nearby Taunton, had in 1565 testified at the trial of William Jayne and Martin Barlowe, merchants of Bristol, both accused of dealing in pirate's goods; but unlike various other sea captains of Frobisher, he does not seem to have indulged in piracy. At the time of court proceedings he was 34 years old and, therefore, during the third voyage 47, three years older than Frobisher. By 1564 he owned his own ship and by 1574 had traded in goods from the tropics, which he had probably acquired on location. At the time of the third voyage Newton was a hardened sea captain, experienced in overseas travels and well qualified for service in the expedition.[1]

The Emanuel in the third voyage

The *Emanuel* had a minimum ships' company of 12 (Hogarth 1993b). Her captain and owner was Richard Newton, her master, John Smythe. There were seven miners, a small crew (of which we know two names, James Leeche, probably master's mate, and Thomas Hancoke, a sailor) and at least one passenger (Thomas Wiars). Wiars, first assigned as boatswain to an un-named, commissioned ship, failed his physical and was discharged-"being found unfytt for service" and was taken aboard the *Emanuel* as one of the unpaid company. Smythe likewise was taken aboard after failing his medical

Figure 11: A small (40-ton) buss, probably resembling the larger *Emanuel*. Drawing by Jan Porcellis, engraving by C.J. Visscher, published (*c*.1632) by G. Valk. Print A9440, National Maritime Museum. Published with permission.

examination; he had been earmarked as master of the *Michael*. The *Emanuel*, with other vessels of Frobisher's fleet, assembled at Harwich on May 27, 1578 and set sail four days later.

During the voyage, the *Emanuel* suffered particularly strenuous trials near the mouth of Frobisher Bay. The first major test took place on July 2, a day that began with promise of sun and calm, but before noon "fayre changed to fowle with thick fogge and boysterous wynde w[hi]ch inclosed and buffited them so on every side that, to defend themselves from the beating of ye Ise, they wer[e] forced to hang over cabels, planks, Ankerstoks and great peases of masts in rope hausers". All hands were busy fending off the driving ice with pikes and oars. The storm continued into the next day.[2]

On July 26, a second storm built up from the northeast and soon became a howling gale driving a blinding snowstorm. The snow froze to the planking and by the evening it lay 6 inches deep on the hatches. In the morning the wind had abated and some ships were frozen in, but "ere it was 12 no[o]ne the [newly formed] yse was gone".[3]

In this storm, the *Emanuel* was hemmed in by icebergs and had a particularly strenuous ordeal. On July 27 she escaped from the ice and met with the *Anne Francis* whose sailors reported:

> their Shyppe to be so leak[i]e, that they muste of necessitie seeke harborowe, having their stem so beaten within theyr huddings, that they hadde much adoe to keepe themselves above water. They had (as they say) five hundreth strokes at the poupe in lesse than halfe a watche, being scarce twoo houres; their menne being so over-wearied therewith, and with the former dangers, that they desired helpe of men from the other Shippes (G. Best in *SM* **1**: 97).

The third storm took place in early September, when the *Emanuel*, *Gabriel* and *Michael* were busy loading at Beare Sound. After a minor reshuffle of personnel, John Smythe found himself in the *Ayde* and James Leeche became master of the buss. The morning of September 1 began with fair weather: there was no wind but an ominous high swell rolled in from the northwest. That evening, Captain Best and his company were accommodated aboard the *Emanuel*, drawing on her provisions, which must have been severely depleted by the end of the voyage. By the morning of September 2 the storm had built up and Frobisher went ashore by pinnace to hasten the loading. The *Gabriel* and *Michael* were filled but, by the time the *Emanuel* was full, the storm blasted down in all its fury. Frobisher was unable to reach his own ship and was compelled to return to England in the *Gabriel*. In the meantime, the *Emanuel* was left in the lee against the rocks and finally had to steer northward into Cyrus Field Bay via Lupton Channel. "The thirde of September being fayre weather, and the wind North northwest she set sayle and departed thence". September 8 they "fell with Frisland" (probably Cape Desolation, Greenland), September 12 "50 leagues" south - by - southeast "descryed" a sizeable

island (probably Cape Farewell),[4] and September 25 sighted the "west part of Ireland
about *Galway*" (T. Wiars in Hakluyt 1589: 635; *SM* 2: 253). On board were 110 tons
of rock from three mines (Table 4).

Fort Dún-an-Óir

Emanuel's unscheduled port-of-disembarkation was a fort later known as Dún-an-Óir,
near the western extremity of County Kerry, Ireland (Hogarth 1989). It seems
appropriate to devote a few lines to the situation here at the time of the *Emanuel*'s
arrival, in order to depict the setting.

In the 1570s, the site of Dún-an-Óir was occupied by one Piers Rice, a citizen of the
village of Dingle and a member of a well established family in southwest Ireland. He
had rented from Gerald [Gerot] Fitzgerald, the 15th Earl of Desmond, the farm or
carucate of Ardcanny, on the west side of Smerwick Harbour, Co. Kerry, comprising
a shragh of 20 acres under tillage, at the annual rate of 13s 4d English money.

Rice appears to have established himself on Dún-an-Óir peninsula, where he built a
"perty castell" and traded produce with visitors to the harbour. Gerald levied a duty or
marte on merchandise imported and exported, charged against Rice in "good fat cows"
or their equivalent value. A further duty was levied against "every boat coming thither
to fish in the harbour".[5]

Into this environment drifted Richard Newton and his *Emanuel*. They were hardpressed
to find safe haven, having endured the rigours of three violent storms in the Canadian
Arctic and possibly others on the return voyage. The *Emanuel* had sighted Galway, on
September 25. This three-week journey from Baffin Island seems to have been one of
normal duration for a trans-Atlantic crossing at this time, and further storm damage may
have taken place off the coast of Ireland. In any event, Gerald Desmond, in a letter to
Privy Council, reported:[6]

> Richard Newton of Bridgewater came w[i]th a Shipp of his called the Emanuell
> of Bridgewater and one of the fleete of Meta incognita laden w[i]th golde eyore
> [ore] to the weast Coaste of the Realme by force of contrarie windes and fowle
> Weather wheare it was tossed up and downe [the west coast of Ireland during]
> the space of sixe or seven weeks in hope of attaine the Coaste of Englande. But
> all happening contrary [she] hath be[e]n dryven in to the Baye of Smarwicke
> w[i]thowte masts, sailes, boate, ancre or eny other convenient furniture ... and
> throughe the misfortunes aforesaid, was driven to discharge the eyure saving an
> eight tonns of great stones w[hi]ch remaine in the ship, and for unlading of the
> quantitie discharged, he was faine to hier two pinasses of the dingell [Dingle
> Peninsula] and hath put the same into a forte of one Piers Rice, and left two of

his men together with a servant of the said Rice to attend the same for the more safetie as well of it, as of the ship w[hi]ch is la[i]d upp to the full Sea marke, and he hath also taken order, that if the ship be not able to contynue, to save the eyore w[hi]ch is aboorde, that the same shall be putt in safetie...

From this account, it is evident that the *Emanuel* had been beached as a wreck near Dún-an-Óir, construction of the fort had already begun and about 100 tons of Newton's ore transferred to the premises, whereas 8 tons of ore were left in the ship.

How and when Newton and his men got back to England is not known. Payment on March 10 1579, presumably in England, to his seven miners for seven months wages is recorded in the expenses of Thomas Allen and Frobisher for the third voyage.[7] Payment to the master, John Smythe, is recorded by Sellman.[8] Thomas Hancoke, a sailor (the only one noted), was paid a month's wages by Frobisher.[9] There is no positive evidence for deaths on the *Emanuel* during the voyage, although, in the case of three miners, direct payment was not made (and could have been transferred to the heirs).[10] However, as no data on the crew are given in the Exchequer accounts, and considering the ordeals of weather and starvation, it is quite possible that the discrepancy in total deaths in the third voyage, recorded by Sellman (13 in *SM* 2: 65-73) and inferred by Best (40 in *SM* 1: 122), can be at least partly explained by the loss in the *Emanuel*.

Unlike the non-commissioned ships whose ore was unloaded at Dartford, there was no allowance for expenses of the *Emanuel*. Newton, who must have paid his sailors from his pocket, was not compensated, and Frobisher, who had paid part wages of three men, was not reimbursed. The commissioners were adamant "the manuell brydgwatter, it is no frayghted ship in Irland" and no compensation could be made.[9]

Newton then sought help from the Privy Council. He, along with his ship and crew, had been pressed into service by Frobisher, under authority of Royal Patent.[11] Could he now take possession of the ore? Newton had insisted that his cargo be kept safe within the fort at Smerwick and Gerald Desmond vouched he would "preserve the premises [i.e. the ore] for her ma[jes]tie ... [un]till better order be taken".[6] The transfer of ownership to Newton was ratified by Act-of-Council, March 25, 1579, and Desmond was ordered to release the ore immediately.[12] However, considering the difficulty of trans-shipping heavy material from the rock-studded and normally turbulent shallows off Dún-an-Óir, plus the fact that it soon became common knowledge that the ore had no value, it is unlikely that Newton ever took delivery.

Fortification, occupation, and siege at Smerwick

The history of Fort Dún-an-Óir is connected with the Frobisher voyages, both in personnel and materials. The first occupation (July 1579) was by James Fitzmaurice

Fitzgerald with a small force of Spanish, French and Italian soldiers. They quickly took over the premises from Piers Rice and three or four youthful servants. According to the contemporary Philip O'Sullivan Bear, Fitzmaurice "fortified it during six days of continuous work. Moreover, on the mainland in front of the rock he constructed a trench and mound, and stationed there cannons taken from the ships" (Byrne 1903: 21). The purpose of this fortification was to establish a beach-head for an invasion of the British Isles by a later, Spanish force.

Fitzmaurice returned three of his ships and about 200 men to Spain, retaining three smaller ships and about 80 Spaniards, Frenchmen and Italians, with a handful of Irish and English soldiers (numbers vary in the several accounts). Sir William Drury, the newly appointed Lord Justice of Ireland, declared open season on Fitzmaurice and his adherents: "if any Captens whatsoever [encounter their ships], then be it that you bringe those vessels and men into the nexte Convenyent harbor of this Realme, there to be kept till order shall be taken".[13] News of the landing of Fitzmaurice was conveyed to Captain Thomas Courtenay [Courtneie] via Henry Davells, a special constable in Dingle village, 17 km (10 miles) southeast of Dún-an-Óir. On the very same day as Drury's detailed proclamation (July 29), Courtenay left Dingle harbour "and having a good wind, did come about and doubled the point [Slea Head], and came into the baie of Saint Marie weeke or Smerweek, and [found there] three ships of James Fitzmoris at anchor" (Holinshed 1587: 154). At the time, the fort was virtually deserted, the soldiers being occupied nearby in a skirmish with the local Irish. Gerald Desmond recounts the naval action as follows:[14]

> Uppon the xxixth of Julie, as the traitor [Fitzmaurice] having the aide of two hondred of the Flaherties that came to his aide by water, weare skirmishing w[i]th some of my men, [when] sodainly came into the haven Captaine Courtney w[i]th a little shipp and a Pynnace, and without any resistaunce toke the Traitors shipps, saving one Barque that he broughte und[e]r the Forte, wher she was broken, so as then the Gallyes of the Flaherties being ronne awaie, the Traitor was like in shorte tyme to Starve within the Forte, or ells to yealde him self to her ma[je]st[ie]s mercye.

Courtenay thereby complied with Drury's request (albeit prematurely), which was said to reflect the sentiments of the Queen.

Cleaveland (1735: Pt. 3, 292) equates this Courtenay with Thomas, son of Sir William, third Lord of Powderham Castle. This was the very same Thomas Courtenay who captained the *Armonell* of Exmouth, a non-commissioned ship in Frobisher's third voyage. In Ireland, he appears to have been paid out of the budget of the Department of Justice and, after his service at Smerwick, was given £400 for his ship,[15] £532 12s 8 3/4d for his services,[16] and a protection with royal pardon.[17]

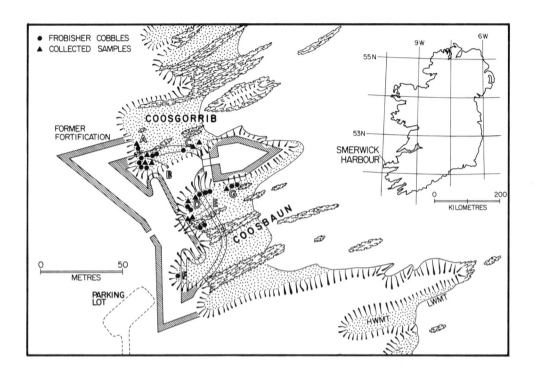

Figure 12: Location of specimens of 'black ore' with relation to the old fortifications at Dún-an-Óir, County Kerry, Ireland. The outline of the coast and fort have been drawn from information in topographic maps, illustrations in Jones (1954), Gowan (1979: Fig. 76) and from personal observations.

Fitzmaurice must have been seriously handicapped through the loss of his principal means of access to and from the fort. In order to avoid starvation, he and his company moved inland and once more the property was left to Piers Rice. On August 18 he was killed in a skirmish with the Limerick Burkes (Webb 1878: 200-1).

There followed a year of serenity at Dún-an-Óir but the government had intelligence of a second, impending occupation. In June, Sir Nicholas White, Master of the Rolls of Ireland, was sent to Smerwick to evaluate the military importance of the fort. He arrived at Dún-an-Óir, June 21, in company with Sir William Pelham, a minor venturer in the Cathay Company and a commissioner, Edward Fenton, Frobisher's lieutenant and a captain in the second and third voyages, and Sir William Winter, who had been a venturer from the start, had provided land in London for Frobisher's assay furnaces, and was the principal commissioner of the Company. Pelham, Fenton and White each submitted official reports[18] of the visit but, of these, only that of White makes reference to Newton's arrival and Frobisher's ore:

> At that instant tyme [late 1578] ashipp laden w[i]th m[aste]r Captain furbusshers newe found Ryches hadpened to presse upon the sandes near that place [Dún-an-Óir], whose carcas and stones I saw lye ther.

This report is of special interest in that it notes that the *Emanuel* was reduced to a skeleton at the time of viewing, and that pieces of ore were visible inside. Presumably White made these observations during his tour of inspection on June 21 and, if so, this is the last authentic report of sighting of the ship and its ore.

The second military occupation of the fort began that summer and fall, on the arrival of four ships carrying several hundred Spanish soldiers under command of Sebastiano di San Guiseppe. Some were also transported from Biscay by none other than Henry Carew, Frobisher's erstwhile captain. They immediately proceeded to strengthen Dún-an-Óir, but at Lord Grey's arrival (in November) fortifications were incomplete because the Irish expediter was "so slow and negligent in providing timber and mortar".[19]

Before the siege, the fort was sketched twice, once by the English,[20] once by the Spanish.[21] The fort was sited on a peninsula and landward was protected by two lines of ramparts, which enclosed living quarters, an office and a church. It was separated from the mainland by a trench over which was placed a drawbridge. The English map has appeared in a number of recent publications (e.g. a coloured reprint in Rodriguez-Salgado *et al.* 1988: 74). An interesting reconstructed, bird's-eye view was made by Fitzgerald (1855). The Spanish map was reprinted by Jones (1954). Figure 12 shows the fort and ramparts in approximate outline. Maps of part of the fort are given in Gowan (1979: 269, Figure 76) and Cuppage *et al.* (1986: 426, Figure 235).

The English took their battle stations in early November. Lord Grey's infantry arrived November 9 and camped beside the fort. Admiral Bingham in the *Swiftsure* arrived the same day and anchored in the harbour to the north. The remainder of the fleet appeared November 11. Positioned next to the *Swiftsure* was the *Ayde*, Frobisher's 'admiral' in the second and third voyages. Martin Frobisher was assigned to this squadron but his ship, the *Foresight*, was separated from the others by weather, and he spent November in England (Glasgow & Salisbury 1966).

An interesting incident, pertinent to the present publication, is recounted by Klarwill (1924: 51) in a translation of Fugger correspondence of November 19, 1580: "A large ship from Spain laden with grain was forced by a ship of the Royal Fleet to run on the rocks. The crew took refuge in the fortress, but the corn remained in the ship". Glasgow & Salisbury (1966) equate this vessel with 'The broken Spannyard', pictured as a wreck in Coosbaun on the English map of 1580. Another, though perhaps less likely alternative, is that 'The broken Spannyard' and the *Emanuel* were one and the same, and that the latter was simply misidentified by Sir William Winter's son (Hogarth 1989). Further discussion of these alternatives will be given below under "Dún-an-Óir today".

The end of the fort is well known history. The task of reducing Dún-an-Óir befell Lord Grey and Sir William Winter. The siege began November 7 and ended three days later with surrender of the fort. Subsequently, over five hundred occupants of the fort, Italians, Spanish, Irish and at least one Englishman, were killed, only twenty were ransomed. Documents available in the last century are summarized by Bagwell (1890 **3**: 13-22, 65-78) and further information that has appeared since has not changed the general picture.

Dún-an-Óir today

Only traces of the original fortification remain. The survey of Gowan (1979: 270-2) shows an inner rampart surviving to a maximum height of 1.5 m. The remains of two bastions are evident, although the northern bastion has been almost completely destroyed, through depredation of farming. The fort was defended by a ditch, now 3 m deep and partially earth-filled. At the time of completion, it could not have been much deeper, as indicated by the cliff exposures on the north and south; it did not penetrate bedrock.

Field observations by one of us (D.D.H.) in the summers of 1987 and 1988 have revealed further information. The cut-away side of the inner rampart at Coosgorrib suggests the wall was mainly composed of sand and turf, interspersed with a few reddish blocks of sandstone and conglomerate. Rare, medium-sized, black blocks, fronting locality A, Figure 12, may represent Frobisher's 'black ore'. They were inaccessible and could best be observed with binoculars.

Figure 13: 'Black ore' at Dún-an-Óir. a = angular block at location C Coosgorrib; b = cobbles at location E Coosbaun.

Figure 14: Coosgorrib in 1989. 'Black ore' cobbles were found on the beach immediately below the 20 - m. cliff. Above the cliff is the remains of the inner rampart and bastion of Fort Dún-an-Óir.

On the beach below, intermingled with shingle of local formations, were a few black cobbles and boulders of possible 'black ore'. One of these, at Coosgorrib (locality C, Fig. 12), was firmly emplaced and could not be budged; it measured 28 x 21 cm on the exposed surface and was at least 14 cm deep. Boulders of similar size and type were seen at D and E, Coosbaun (Fig. 12). The largest boulder, 45 x 35 cm and more than 15 cm deep, was found at G. Also notable was the occurrence of angular black specimens, commonly associated with well rounded ones (Fig. 13).

Finally, the distribution of the black cobbles follows definite patterns. Most are within 2 m of the closest cliff face and never more than 5 m from it. At Coosbaun, they are close to the site of the inner rampart and not south of it; at Coosgorrib they are not found north of the outer rampart. They tend to be in clusters, labelled A to F in Figure 12; some are associated with small strands of black sand.

All of these facts are consistent with the hypothesis that Frobisher's 'black ore' is incorporated in the ramparts protecting the fort, and perhaps the revetment, which periodically tumbles to the beach below. That the rock represents the *Emanuel*'s cargo, appears to be correct, but the hypothesis that it forms the remains of the '100-ton trove', piled on the beach by Newton (Hogarth 1989), must now be discarded. Certainly the force of waves in Coosgorrib (the very name means rough cove) and violent sand washing suggest that it is improbable that any block could have retained a large size and angular shape on the beach for one year, much less four hundred. In other words Hogarth was led to the correct locality by following an incorrect theory: it was assumed that these rocks had, in 1578, been placed *deliberately* on the beach, immediately *below* the fort.

As noted above, the identity of 'The broken Spannyard' remains uncertain. It may well represent the grounded Spanish victualler, as suggested by Glasgow and Salisbury (1966). However, a concentration of black boulders near G, Figure 11, is in the approximate position of The broken Spannyard as marked on the 1580 English map, but the proximity of this concentration to the rampart may be more than fortuitous and its persistence through the centuries tends to refute any theory connecting its presence here with the *Emanuel*'s last resting place.

Artifacts located with an underwater metal detector were itemized by Snoddy (1972). Some of these, such as a 16th century buckle, a lead ingot, and fragments of brass cannon, might belong to the period of the siege of Dún-an-Óir (1580), or the occupation of Fitzmaurice (1579), but they could equally well be ascribed to the *Emanuel*. It should be kept in mind that Newton's vessels were armed.[22] Albeit, whoever continues the underwater search will have his, or her work cut out: Smerwick Harbour appears to have been the graveyard for ships, with continual additions, for more than four centuries.

Notes

1. Most information in this paragraph has been taken from R. Newton's deposition, June 7, 1565. *PRO*, E 159/350, Easter 418 [in Latin], and Water Bailiff Accounts *SRO*: D/B/bw, 1469 (1572) and D/B/bw, 1470 (1574).

2. Information and quotation taken from the hitherto unpublished, anonymous account of the *Judith*, 1578. *BL*, Harley MS 167/41, f.182R.

3. C. Hall, Ship's Log of the *Thomas Allen*, completed Sept. 28, 1578. *BL*, Harley 167/42, f.192R. According to Thomas Ellis, a sailor thought to have been on the *Ayde*, one foot of snow covered his boat (*SM 2*, 41).

4. Confusion has resulted from placing Cape Farewell in 57½ N latitude, 2½ degrees south of its true position, and poorly defining the coast of Greenland ['Frisland']. This led seventeenth and eighteenth century geographers to site the island ('Buss island') in the Atlantic deep (see Babcock 1922, Christy 1897 and Johnson 1942). Best (in *SM 1*: 122) muddied the waters further by describing the land as "fruitefull of woods, and a champion countrie".

5. Anon. *A survey of the honours, manors, lordships, lands etc. forfeited by Gerald, 15th Earl of Desmond, and his adherents, in Kerry*, 1584. MS, *RIA* [Translated from the Latin original, in the Public Record Office, Dublin]. See also Inquisition of 1584, *Kerry Archealogical Magazine*, **1**, 217-8, 266 (1910).

6. Desmond to Privy Council, Jan. 9, 1579. *PRO*, SP 63/65/15.

7. The Account of E. Sellman, Mar. 20, 1579. *PRO*, E 164/36, f.133. 'Reddy monneye paid by master Allen the 10 marche'.

8. Ibid., f.152. There is no mention of James Leeche in the accounts, who was the master according to Thomas Wiars (*SM 2*, 253).

9. Account of Martin Frobisher, Jan. 21, 1579, *PRO*, E 164/35, f. 290. Auditors' summary; 'parcells nott allowable'.

10. Account of Thomas Allen, Mar. 13, 1578. *PRO*, E 164/35, f.259. Wages of Robert Pemberton, John Hetherington and John Brown.

11. *Calendar of Patent Rolls, Elizabeth I*, **7** (**2886**). Mar. 21, 1578.

12. *Acts of the Privy Council*, N.S., **11**:60, Feb. 26, 1579; *ibid*.: 89, Mar. 25, 1579.

13. W. Drury to all vice-admirals, captains by sea, and masters of ships, July 29, 1579. *PRO*, SP 63/67/65. Similar but less explicit proclamations were issued by Drury on July 24, 1579. *PRO*, SP 63/67/47 and 48.

14. Proclamation of G. Desmond, Oct. 10, 1579. *PRO*, SP 63/69/51.

15. Sir H. Wallop to Lord Burghley, Oct. 17, 1579. *PRO*, SP 63/69/68.

16. Sir H. Wallop to Lord Burghley, Nov. 29, 1579. *PRO*, SP 63/70/38.

17. Pardon and protection to Thomas Courtenay, Feb. 16, 1580. *PRO*, C 66/1188, m. 5 [in Latin]; Calendar of Patent Rolls, Elizabeth I, 1578-80, No. 1496.

18. Sir W. Pelham's journal, June 21, 1578. *ALL*, Carew MS 597, ff.62R, 354R. E. Fenton to Sir F. Walsingham, July 11, 1580. *PRO*, SP 63/74/21. N. White to Lord Burghley, July 22, 1580. *PRO*, SP 63/74/56.

19. Sebastiano di San Giuseppe to the Cardinal of Como, Dec. 31, 1580, quoted by O'Rahilly (1937: 9). Spanish advertisements, June 23, 1580. *PRO*, SP 94/1/51.

20. Map by William Winter, Jr., Dec. 24, 1580. Map - *PRO*, MPF 75; accompanying text - *PRO*, SP 63/79/103.

21. Map by Angelo Angelucci, Sept. 29, 1580. Map - *ASV*, Nunziatura di Spagna **25**, ff.375V - 376R; accompanying text by Sebastiano di San Guiseppi [Bastian di San Joseppi], Oct. 1, 1580. *Ibid.*, ff.370R - 374R.

22. Water Bailiff Account, 1572. *SRO*, D/B/bw, No. 1469; *ibid.*, 1574. *SRO*, D/B/bw, No. 1470.

CHAPTER 4

Proofs and Furnaces

The metallurgy of the Cathay Company progressed in a logical manner from small to large proofs and then to pilot plant production. Paramount in this development were the large proofs made in London to assess the grade of the 160 tons of 'black ore' of the second voyage. Some of the samples weighed 2 cwt (Table 5). Unfortunately, quantitative data are lacking for tests that gave low values, results considered to reflect failures in technology. Contemporaries blamed Frobisher, insinuating that he selected his men at random or through favouritism or nepotism, thereby acquiring a staff of incompetents. By mid-1578, even Michael Lok had turned against him and, what is probably the best account of the development of the Company in its later stages, is to be found in Lok's scathing criticism *the doynges of Captayne Furbusher*.[1]

In this chapter we will consider assays of Frobisher's ore and the furnaces set up to treat it (Table 5). Tests 1 to 6 and 12 to 14 were made from furnaces in London and suburbs; tests 7 to 11 were made at Dartford. Assays 1, 2 and 6 were termed by Lok the "first, second and third great proofs".[2]

Jonas Shutz and the London assays

Jonas Shutz, assayer and servant of the Duke of Saxony, came to England on temporary leave-of-absence. From March to November 1576 we find him in London and Keswick with the Company of Mines Royal, possibly representing the major shareholder, the firm of David Haug, Hans Langnauer & Co. of Augsburg, Bavaria, and attempting to upgrade their faltering copper smelter in the Lake District (Collingwood 1912: 184, 186, 189). Shutz had assisted Giovanni-Battista Agnello in the very first assays of ore from the Baffin Island area. He had independently made small proofs and, whereas certain reputable assayers had failed to recover precious metals from Frobisher's rocks, Shutz had, in every assay, managed to find at least some gold. Considering the aura of respect with which Saxon metallurgists were held in Elizabethan times and Shutz' success in wringing gold from an ore that was extremely difficult to work, it is not surprising that he won the admiration and confidence of commissioners and venturers alike. Shutz blamed the failure of others, as well as those of his own, on the tenacious and refractive nature of the ore, "the gold remayninge in the slags w[hi]ch coulde not be well brought outt"[3] and the available furnaces being incapable of attaining sufficiently high temperatures. The credulous public believed him. With such esteem, he was the natural person to be appointed principal assayer on the second voyage and, later, chief assayer of works in London and Dartford. He was essential to the progress of the Company and, at the beginning of 1578, a critical period in the Company's history, we find Lok pleading with Queen Elizabeth to extend his sojourn in England.[4] Apparently, this request was successful.

Table 5: *Bulk assays in London and vicinity and pilot-plant tests near Dartford*

No.	Date completed	Wt (cwt)	Location	Assayer	Voy	Assay (oz/T) Au	Ag
1	Nov. 1, 1577	1	Tower Hill	J. Shutz	2	13.3	
2	Dec. 6, 1577	1	Tower Hill	J. Shutz	2	13.3	
3	Jan. 23, 1578	1	Agnello's house	J-B. Agnello	2		
4	Jan. 30, 1578	2	Cripplegate	J. Broad	2		
5	Feb. 21, 1578	1	Kranich's house	B. Kranich	2	13.5	51
6	Mar. 6, 1578	2	Tower Hill	J. Shutz *et al.*	2	2.6	64
7	Nov. 13, 1578	40	Dartford	J. Shutz	2		
8	Nov. 13, 1578	40	Dartford	J. Shutz	2		
9	Dec. 29, 1578	20	Dartford	J. Shutz	2		
10	Jan. 20, 1579	10	Dartford	J. Shutz	3		
11	Feb. 17, 1579	10	Dartford	J. Shutz	3	1.2	35
12	Mar. 24, 1579	2	Tower Hill	J. Shutz	2	1.7	49
13	July, 28, 1583	1	Tower of London	W. Williams	2(?)	nil	0.04
14	July, 28, 1583	1	Tower of London	W. Williams	2(?)	nil	0.1

Abbreviations: *BL* British Library, Lansdowne MS 100/1 [*c.* 1581]; *H&L* Hogarth & Loop (1986), pp. 260 (with number of assay) and 261; *PRO* Public Record Office; *SM* Stefansson & McCaskill (1938), vol. 2.; Voy Voyage represented.

(table 5, cont.)

No.	Other data and remarks	PRO	SM (page)	References BL (folio)	H&L (page)
1	'First great proof'	SP 12/122/62	124	4R	260(1)
2	'Second great proof'	SP 12/122/62	125	4R	
3	"yt succeeded not well"	SP 12/122/62	126		
4	"It did succead well"	SP 12/122/62	126		
5	Silver parted with stibnite	SP 12/122/67	127-9	5V	260(2)
6	'Third great proof'	SP 12/123/5	131-2	4V	260(4)
7	The result proved "verye evill"			10R	
8	Results were "somewhat reasonable"			10R	
9	Work "suceeded but evill"			12V	
10	Assay gave £ 10/T			12V	
11	35 oz/T silver remained in slag	SP 12/129/43	149-50		
12	28 oz/T silver remained in slag	SP 12/130/15	150-1	13R	260(6)
13	Values obtained from bead diameter	SP/12/161/41			261
14	Values obtained from bead diameter	SP/12/161/41			261

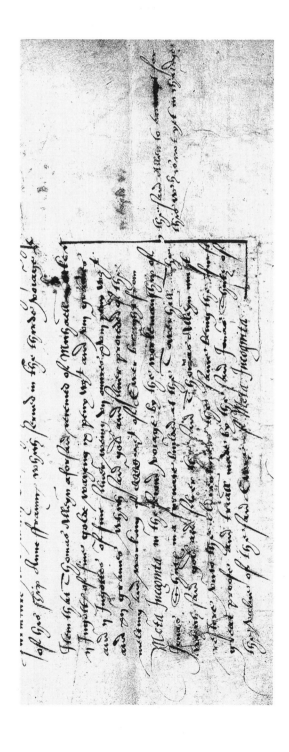

Figure 15 (opposite): Paragraph from *Declaration of the accompte of Thomas Allen*, written on parchment as Chancery Miscellanea [*c.* 1580]. *PRO*, C 47/34/6. Published with the permission of the Public Record Office (U.K.).

Item that Thomas Alleyn aforsaid received of Michaell Locke ij Ingotts of fine golde wayeng ix penywgt and viij graines and ij Ingottes of fine silver weing vij ounces xviij penywgt and xij grains which said gold and silver proceded of the melting and working of CCCC wgt of Ewer, brought from *Meta Incognita* in the second voiage by the workemanship of Jonas Shuts in a furnace buylded at the Tower hill. The which said gold and silver the said Thomas Alleyn must restore unto the Adventurers, the same being the furst great proofe and triall made by the said Jonas Shuts of the value of the said Ewre of *Meta Incognita*.

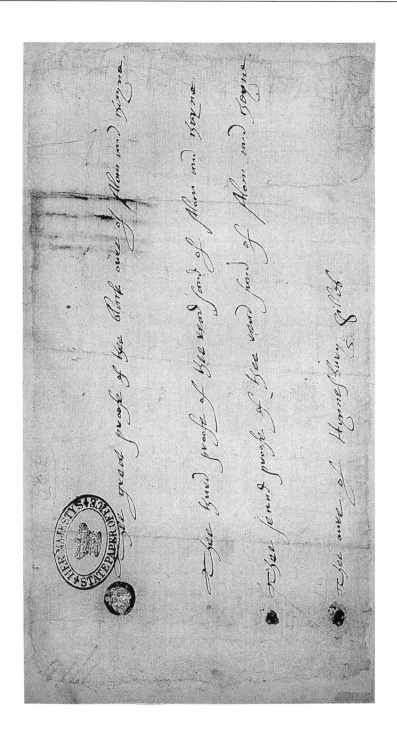

Figure 16 (opposite): Assay sheet endorsed *proofes of Furbusher ewer*.

Thee [first] great proofe of thee black oure of Alom and chayne.
Thee thurd proofe of the read sand of Alom and chayne.
The second proofe of the read sand of Alom and chayne.
Thee our of Hynnesbury hilles.

All assays originally had gold implanted in a dob of red wax on the left side of the sheet, from which a large (top) and small product (bottom) remain. 'The first greate proofe' was made by Jonas Shutz and reported Nov. 1 1577. For "Alom and chayne [chain]" read "alum *et cetera*". The "proofes of the read [red] sand" were made by Burchard Kranich and reported Nov. 26 1577. The document is probably in error in ascribing the same method of parting (nitric acid extraction) to Kranich's test, where silver was normally sulphidized with stibnite. 'Hynnesbury hilles' cannot be located. *PRO*, SP 12/119/9. Published with permission of the Public Record Office (U.K.).

Lok tells us that Shutz began to build his furnaces at Sir William Winter's house on Tower Hill, in October 1577,[5] about the time of return of ships from the second voyage. The plot marked the site of a former chapel and college of priests, the latter dismantled in 1548. It had then been cultivated, and upon its gardens Winter had set up a large storehouse of "strong frame of timber and bricke .... imployed as a fayre house of Marchants goods brought from the sea" (Stow 1598: 95-96). It was well located: just a few hundred metres west of the Tower (where the ore was stored) and an almost equal distance north of the Thames (the principal route of supplies). The two furnaces, 'melting' and 'refining', were modelled by Shutz. In Lok's expense accounts we find the following items, which pertain to the first great proof:[6]

> paid to Olyver Skiner for yron, 1^c.2.12 ^{lb} [1 1/2 cwt, 12 lb],
> at 11th the C[cwt] to make bares [bars] for the Furnaces . . . . li 0.18.0
>
> paid to Jonas Shuttz to buye things necessarye . . . . li 0.10.0
>
> paid to henrike Williames, brickelear, and Robart Denam,
> goldsmithe for ... stuffe and workemanshippe for the
> bulding of ii furnaces . . . . li 11. 0.6
>
> paid for vi sackes & vi dossen basketts to carye the Ewer
> from shippe Michaell ... into the tower . . . . li 0.12.6
>
> paid Clypsam, plomar, for C leade to melt w[i]th Ewer . . . . li 0.12.6
>
> paid to Robart Denam for billetts, bone ashes and other
> necessaryes for the meltinge of Ewar . . . . li 0.18.3
>
> paid to harry gisborne, smythe, for tooles Irone for
> the furnaces . . . . li 1.3.4
>
> paid to John fyshe, smythe, for yron worke and tooles
> for the ii furnaces at tower hill . . . . li 2.11.0

The first great proof, like all succeeding proofs at the Tower was made in a wind furnace, fueled by wood and coal and blown by man-driven bellows. It involved considerable labour, as the following account will show:

> paid to 4 labourars in the towar helpinge to waye the Ewr
> there laden in ii dayes . . . . li 0.10.0
>
> paid for carriage of 15 Lodes Ewar into the Towar
> out of shippe Michaell at 8^d p l [per lode] . . . . li 0.10.0

paid for stampinge Ewar & helpinge to melte at towar hille,
for ii Labourars, 8 dayes . . . . li 0.13.6

paid for ii labourars for 6 dayes to blow bellowes and for
workinge in the meltinge Ewr . . . . li 0.10.6

Unskilled labour was shamefully cheap, but goods were relatively expensive in Elizabethan times!

According to the audit of the Cathay Company, the 'furst great proofe' generated two ingots of pure gold, aggregating about 1/2 ounce, and two ingots of pure silver, aggregating 8 ounces (Fig. 15). These ingots were appropriated by the second treasurer of the Cathay Company, Thomas Allen, and held ransom for backpayment of freight in the voyages.[7]

All three 'great proofs' appear to have been made in Sir William Winter's warehouse,[8] and the furnaces were rebuilt after each run. The third proof was particularly significant, as it produced dishearteningly low values. The test was made in March in the presence of commissioners and, after a preliminary trial, the melting was deemed successful. Jonas Shutz was assayers-in-chief, Humphrey Cole, John Brode and Robert Denham were assistants. They consumed at least 3 1/2 cwt lead, unidentified fluxes ('mixtures' purchased by Denham) and 2 lb nitric acid (to separate gold from silver). The furnaces were blown for a week by two men at foot-bellows; Shutz became seriously ill with smoke inhalation. From 2 cwt of sample, they managed to produce an assay registering but 0.26 oz gold and 6.38 oz silver per ton, a sufficient grade for profit only at over-optimistic cost estimates of mining, transportation and ore-preparation. However, according to Shutz, "much gold remayned in the slegs [slag]".[9] Mindful of the high assay recently completed by Kranich (No. 5, see below), the Commissioners accepted this explanation.

Burchard Kranich and his contributions

Late in 1577 Burchard Kranich enters the picture. He had been responsible for introducing the mechanical jig, water-driven stamps and an improved mine pump to England. He had set up smelters and managed lead-silver operations in Derbyshire and Cornwall.[10] By December 1577 he had already made trials of the original specimen from Little Hall's Island. As for proofs from ore of the second expedition, it was considered advisable to have a second opinion and, on December 10, Kranich was brought into 'conference' with Shutz. It seems antipathy was mutual and almost immediate. Within three or four days Shutz developed an intense dislike of Kranich "boethe for his evell manners and ignorance in divers points of handelynge of the ewer".[11] For his part,

Kranich declared that "yf Jonas had any couninge yt had longe since [dis]appered" and he forbade Shutz or his 'confederats' to approach him in future.[12] It was inevitable that the administration should split into two cliques: in issues concerning the valuation of ore, Shutz sided with Fenton and Lok, Kranich with Frobisher and Winter. Robert Denham, assistant to Shutz on the second voyage, seemed to remain neutral and was a go-between on numerous occasions. Strangely, all three assayers, Shutz, Kranich, and Denham, were awarded patents to manage the melting and refining of the northwest ore.[13] Kranich was secretive and refused to divulge his methodology but he finally permitted a select group to view a large proof near its conclusion. A delegation of commissioners watched him pour his crucibles and obtain Assay No. 5, Table 5. The 1 - cwt sample "was melted in potts w[i]th additaments, by halfe pounds in a pott". This was "not the order of the great workes" [Dartford] and must have been a tedious process.[14] Regardless, Kranich produced an assay with values similar to those of the 'first' and 'second great proofs' (Nos 1 and 2, Table 5) and promised to extract 10 ounces from every ton of like ore in the future. It was upon this assay and promise that the 'foundacyon' for the third voyage was firmly set.[15]

Shutz was critical of Kranich's work and commented caustically "noe man but himself knoweth wate he puts in additaments [additives]". In the meantime, Denham was prevailed upon to obtain some of Kranich's 'antimonye' additive, and on February 22, 1578 it was tested before the commissioners. From it, they found silver at the rate of 30 ounces a ton.[16] Kranich had refused to continue assaying without his 'antimonye', which, after it was shown to be argentiferous, neither commissioners nor Lok were happy to allow. Besides, the 'antimonye' was collected 10 years earlier[17] and the supply in London was fast running out. Thereupon Kranich, who to this point had been chief assayer, was disgraced. But he never divulged the source of his 'antimonye', other than saying it came from the grounds of a Mr Edgcombe, near Saltash, Cornwall.

Whereas Kranich was suspected of salting his assays, neither commissioners nor Lok were completely convinced that 'antimonye' had been debunked. They spent much money and effort in trying to locate its exact source. The Privy Council requested aid from Piers Edgcombe,[18] who did not know the location himself, and Fenton was sent into Cornwall and Devon to try to find it.[19] It was not until January 13, 1579, after the furnaces at Dartford were in full operation, that Fenton was able to announce triumphantly "Mr [Richard] Edgcombe hathe found the like ore delivered to Burcot", but to this point the discovery was based on hearsay, neither of the Edgcombe brothers having seen the vein. Richard requested "if happelie he [should] find the load, what allowance [the Company] woulde give him for the tonne thereof to be delivered at dartford".[20]

It was Kranich who suggested that Sebastian Copland, a German millwright with expertise in smelting, join the staff. He was then employed in the iron works at Grimsthorpe, near Edenham, Lincolnshire, by Richard Bertie, husband of the Duchess

- Dowager of Suffolk.[21] On the request of Sir Francis Walsingham to the Duchess, Copland was released from his former occupation, and brought to Dartford by William Hanke, Frobisher's cousin and servant.[22]

About the same time, Hans Staddeler, a 'docheman' metallurgist and former employee of the Mines Royal in the Lake District, joined the staff. In July 1578, he was sent to Hamburg and returned with four 'docheman' assistants.[23] They arrived in London in time to rebuild the furnaces at Sir William Winter's house and look after the 'third great proof'. Henceforth the furnaces were essentially manned and operated by Germans. Kranich, now plagued by ill-health, fades from the scene. Official notification of his death was released October 22, 1578 (Donald 1950: 320) but Shutz, Copland, Staddeler and his 'dochemen' continue into the Dartford era.

After return of the second voyage, other furnaces were set up in the London area, with expenses paid through the Company. These were at the residences of Giovanni-Battista Agnello, George Wolfe, Burchard Kranich and Martin Frobisher.[24] They were small furnaces, probably operated for minor proofs at time of peak demand. But, by then, the London furnaces had outlived their usefulness and it was time to look into establishing larger works elsewhere.

The Dartford enterprise

In late 1577, the north Kent market town of Dartford, situated on the River Darent and only 30 km (20 miles) from London (Fig. 17) was chosen as the site of smelting works of the Cathay Company. The following year it became a centre of activity during the construction of a mill and smelter to process Frobisher's ore for precious metals. A small portion of the ore was worked there in 1578-9. Later, when it fell from favour, some of the stockpiled ore became incorporated in the wall of the Manor House, part of which still remains intact in the town centre.

The site of the mill and smelter was near the centre of Bignores Manor. Before the dissolution of the monasteries, the estate had been owned by the Dominican nuns of the Dartford Priory. This long-established mill site later became the home of England's first successful paper mill (1585-1732) and the site of the Dartford Gunpowder works (1732-1907). Table 6 summarizes the history of the mills. Further details are given by Boreham (1986a and b), Dartford District Archaeological Group (1986), and Philp (1986). No trace of the smelting works remains, all evidence having been obliterated by subsequent mill developments spanning more than three centuries.

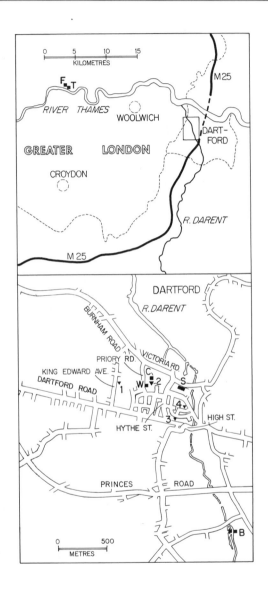

Figure 17: *Upper frame*, location map. Abbreviations: F furnace site (Sir William Winter's storehouse), T Tower of London. *Lower frame*, west-central Dartford. Abbreviations: B mill and metallurgical site (Bignores estate), C ore-storage site (site of chapel, Manor House), S present railroad station, W Tudor boundary wall. Sample locations: 1 King Edward Avenue (sample AC), 2 Henry VIII Manor House (sample series E1, E23, E44), 3 High Street (sample series E43), 4 Bullace Lane (sample series E47).

Table 6: *A chronology of the mills and factories at Bignores*

Year	Lessee (L) or Owner (0)		Type of mill	Comments
1535	Dartford Priory Roger Tusser	(0) (L)	Wheat & malt	Leased from Dartford Priory at £12 per year
1547-80	King Henry VIII - Queen Elizabeth I William Vaughan	(0) (L)	Wheat & malt	Leased at £12 per year
1578-80	*Queen Elizabeth I* *Company of Cathay*	*(0)* *(L)*	*Metallurgical*	*Probably a different site from Vaughan's mills*
1580-5	Queen Elizabeth I	(0)	Wheat & malt	William Death succeeded to lease of Vaughan, who died in 1580
1585-1626	John Spilman, Sr	(0)	Paper	Wheat and malt mills converted at £1,500. First paper mill in England. John Spilman, Sr. died in 1626
1626-41	John Spilman, Jr	(0)	Paper	John Spilman, Jr. died in 1641
1641-70	Browne, Harris & Gill Owner not known	(L)	Paper	The continued operation of the paper mill under this tenancy is uncertain
1670-79	William Blackwell	(0)	Paper	A paper mill was also established at Brooklands, 2 km downstream
1702-24	William Quelch	(0)	Paper	Dartford mills purchased for £22,000
*c.*1725.32	Richard Archer	(0)	Paper	
1732-?	Pike & Edsall	(0)	Gunpowder	Mills built on, or close to, the sites of former paper mills and metallurgical plant
1778-90	Pigou & Andrews	(0)	Gunpowder	Four mills operated, all destroyed by explosion 1790
1791-1840	Frederick Pigou	(0)	Gunpowder	Powder magazine most extensive in England. Suppliers for munitions in Napoleonic wars
1840-50	Pigou & Wilks	(0)	Gunpowder	
1850-1902	Pigou, Wilks & Lawrence	(0)	Gunpowder & guncotton	*c.*1890 guncotton mill built adjacent to the gunpowder plant
1903-11	Curtis & Harvey	(0)	Gunpowder & guncotton	Production intermittent; gunpowder mill closed 1907; guncotton mill closed 1911
1911-1914	Curtis & Harvey	(0)	Gas mantles	

Selection of the mill and smelter site and preparations

The idea of a plant or 'great works' to extract gold and silver commercially must have originated soon after 'black ore' was shown to exist in quantity during the second voyage.   It then became obvious that the small furnaces blown laboriously by man-pedalled bellows and ores crushed strenuously in huge steel mortars would not suffice for large-scale processing. The Company settled on a smelter and a mill, which included bellows and stamps driven by water power. In Shutz' estimates of November 1577 is an item of £400 for mill and furnace construction.[25]

In early December 1577, Lok, Frobisher and Shutz "vewed all the watter mylles neere London" and those at Dartford were considered specially attractive. Prerequisites for an appropriate mill and furnace site included a constant and reliable source of non-tidal water, ample ground space for construction of buildings,and access to roads and sea water to facilitate transport of ore. The mill site in Bignores Manor met all of these conditions. The trio concluded "we thynk that place good for the purpose".[26] The site was on the River Darent, approximately 2 km upstream from Dartford and 5 km from the confluence of the Darent and the Thames (Fig. 17).

In 1577, the Bignores site was occupied with a wheat and corn "mylle under one roffe", under tenancy of William Vaughan. He had leased the farm and mills since August 1547. The prospective long-term loss of the mills to the Cathay Company threatened to deprive Vaughan of his main source of income and he filed an application for compensation for himself and the two tenant millers. Vaughan had invested money in the refurbishment of the mills and pleaded that requisitioning the mills would cause hardship and loss of income "beynge the greatest staye of his lyvinge". He was elderly and had served Elizabeth well. He therefore asked the Queen to extend his lease to 21 years. This would help to off-set the charges he would undoubtedly incur if he had to relocate. He further asked that John Wynde and Richard Clarke, the incumbent millers, be compensated for inconvenience.[27]

Thomas Fludd, the surveyor, examined the Dartford site in the last week of December, 1577. In that time, a surveyor acted as intermediary for both lessor and lessee (or vendor and buyer). In effect, he was a real estate agent as well as diplomat and pacifier. Fludd was accompanied by Nicholas Chancellor,[28] who represented the Company, and had been purser on all three voyages. They found the mills in "resonable good state of rep[ar]acon" but in need of some maintenance.   Fludd and Chancellor were unable to locate Vaughan but interviewed John Wynde and Richard Clarke, the tenant millers, as well as William Death, co-lessor, farmer and Vaughan's son-in-law. Wynde's tenure was due to expire midsummer 1578 but Clarke's lease was to run another 13 years. Recommendations were embodied in a letter of Thomas Fludd to Lord Burghley of January 7, 1578.[29] It seems that the metallurgical plant was established on a new location and its presence did not affect the old mills, except that, during construction, it was necessary to divert the Darent for four weeks in order to build a platform for the workhouses and, during the interval, the mills of Harbord More (successor to John Wynde?) and Richard Clarke were left high and dry. More and Clarke collected damages of £1 and £6 13s 4d, respectively.[30] The association of the old and new tenants appears to have been amicable and we find Clarke, Wynde, More and Vaughan each supplying material and performing services for the Company during the construction period. Vaughan had his lease extended.[31]

The first official reference to Dartford works in land papers appears to have been a lease in reversion[31] signed by Thomas Fludd (Jan. 28, 1578), countersigned by John Thomson (Feb. 1, 1578), and ratified by Lord Burghley (n.d.). The lease was made out to "W[illia]m Vaughan, tenant, and W[illia]m De[a]the, his son in law, for xxj years w[i]thout fyne, in Considerat[io]n yt [that] they shall partly grant to 1. [The] L[ord] Tre[a]sorer [Burghley] and S[i]r Walt[er] Mildmay, 2. ye Q[ueen's] Ma[jes]ty, use for their Interest of ij water milles w[i]th the appurten[an]ces [and] works". Burghley and Mildmay probably represented the ever-changing slate of commissioners of the Cathay Company.

A directive authorizing the building of a small smelter and refinery at Dartford was passed by the Privy Council, January 8, 1578.[32] January 12 was the date of a detailed examination of the mill site by Frobisher and his chiefs of staff: Jonas Shutz - the metallurgist, Henrik Williams - the stone mason, Sebastian Copland - the millwright, Robert Hadlowe - the master carpenter, and Thomas Hitchcock - the house carpenter.[33] They "measured oute the platt of grownd for errection of the buildings and furnacs". Estimates (still greatly underrated) were itemized and sent to Sir Francis Walsingham, January 19.[34]   Then, the first construction materials began to arrive: freestone (homogeneneous compact limestone) temporarily stored at Muscovy House, London in February, followed by timber,[35] leather for bellows, and other material.

One of the most pressing problems was to find a storage site for the ore before it reached the mill. Fortunately, the Manor House built by Henry VIII at Dartford in 1542 - later the home of Anne of Cleves - was still in sufficiently good repair to house the rock. In 1578, the Manor House belonged to Henry VIII's daughter, Elizabeth I.   Thomas Asheley was ordered by Privy Council to prepare "the olde emptie chappell in her Majesties howse there" for storage of ore. The 'olde emptie chappell' was probably the Queen's chapel near the northeast end of the compound (Boreham 1991: 14), whichexplains the concentration of specimens of 'black ore' immediately to the south (locality 2, Fig. 17). Asheley was instructed "to permitte suche as shall have the charg of the melting and keping thereof to prepare and make readie the place, and to bestowe the said oare therein at such tyme as it shalbe brought thether".[36]

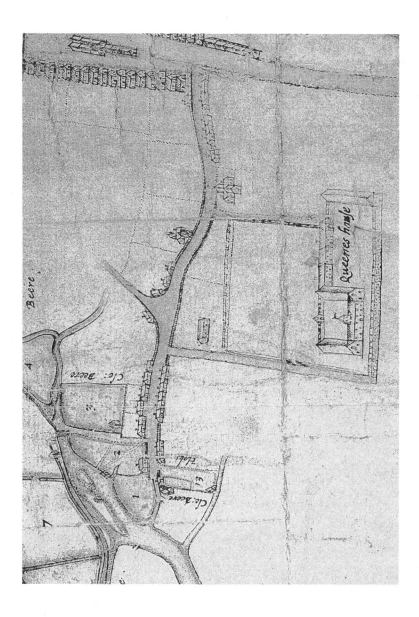

Figure 18: Estate map of 1596 showing land ownership in central Dartford. The Queen's House (King Henry VIII Manor House) occupied the site of the former Dartford Priory. The ore was probably stored below the tower in the upper left-hand (northeast) side of the enclosure. Reproduced by permission of Rochester Bridge Trust, Rochester, Kent.

Figure 19: Burchard Kranich's preliminary sketch of the Dartford furnaces. On the left is a large melting furnace, labelled 'Burcharts furnice', where precious metals, copper and zinc were separated from slag. It was blown by four pairs of water-driven bellows. Three small piles of 'black ore' are in the foreground. To the right is a refining furnace, where gold and silver were separated from copper and zinc. The small furnace behind, driven by two pairs of man-powered bellows, is labelled 'old-order', and may, for comparison, represent a typical London furnace. Both inscriptions are in Lord Burghley's hand. *PRO*, MPF 304. Published with permission of the Public Record Office (U.K.).

Construction of the mill and transfer of ore

Preparations of the mill site proceeded at breathtaking speed in order to have a working mill available to receive ore from the third voyage. Construction began April 23, 1578 and principally involved preparing a platform for office, mill and smelter houses, cutting a sluiceway for water power, building a dam, temporarily diverting the course of the River Darent, as well as setting up the buildings and furnaces. Frameworks for the buildings were mainly oak, the walls were brick and the roofs tile. Edward Castelin was overall supervisor, Thomas Kennyon looked after timber imports, Robert Hadlowe carpentry, and Henrik Williams brickwork. At peak activity, the payroll included 80 carpenters, bricklayers and labourers. Work on the furnaces commenced August 24 but, by September 6, external work was complete and the main construction force was laid off,[37] five months after commencement, and barely one month before the arrival of Frobisher's fleet in the Thames estuary. The final cost of the buildings (exclusive of the furnaces and their maintenance) was £1879 3s[38], more than four times the original estimate.

We know rather little of the mill itself. The first notice, presumably by Shutz, called for a building 84 feet long and 36 feet wide, a stamp mill, melting and 'fining' [refining] furnaces, water wheels, and 3 pairs of bellows.[39] The sketch by Burchard Kranich (Fig. 19) shows the main melting house with a rectangular, top-loading, firebrick furnace, about 6 feet high, blasted with 4 pairs of water-driven bellows.[40] Kranich had complained that Shutz' bellows were placed far too high and would not blow onto the hole of the forehearth. Shutz countered that Kranich's bellows were higher still,[41] an argument consistent with Kranich's sketch. A smaller, circular, melting furnace was blasted with two pairs of manual bellows and may represent a London furnace. A rectangular, refining furnace, did not require blowing. These preliminary plans were later modified and amplified by both Shutz and Kranich, although no diagrams have survived. There are, however, statements in the expense accounts showing that the final plant included ten pairs of great leather bellows, twelve feet long, at least three pairs of smaller bellows, stamps (mainly iron, two brass), axle-trees, and fifteen brass cogs.[42] There were also "two great workhouses, two water mylles w[i]th fyve great meltinge Furnaces [elsewhere four melting furnaces, one refining furnace] in the same houses, and one great Cole house & other necessary workhouses".[43] There were no indications of manually operated bellows.

In the meantime, ore from the second voyage had been transferred from its temporary storage in Bristol castle to the Dartford Manor house. A directive had been issued by Privy Council in early July which gave authority for this transfer.[44] Soon after, 121 tons were loaded into Robert Dudley's *Whitbeare* and shipped to the Dartford wharf, whence they were carted to the old chapel of the former Dartford priory, near tidewater and two kilometers downstream from the mill site. At the same time, 20 tons of rock were moved overland. On both occasions Thomas Marshe, who had been 'merchant' in

Frobisher's second voyage, was expediter. These transfers were expensive: the total charge against the Company was £245 5s 1d.[45] Ore from the second voyage, stored in London, appears to have remained in the Tower, available for later tests of the furnaces in Sir William Winter's warehouse.

Ships of the third voyage anchored in the Thames estuary during October 1578. There was no rush to unload, as the furnace additive ['additament'] had not arrived and even as late as December there were complaints that shipments of 'additament' were in arrears.[46] Most of the lead, litharge, galena and marcasite for the furnace charge arrived in November,[47] and only then did the community spring to life, employing numerous small carts to move the ore southward along the narrow roads. From November 22 to December 18, a total of 1572 loads of ore were transported, from dock to depot, at 3d and 4d the cart, 1218 loads on November 28 alone.[48] Possibly some of these loads may have been 'additament'. Thence a small portion (17 tons) of the ore was transferred to the mill at 10d the cart. Records show 38 loads for this leg.[49] Ore from the second voyage (all from the Countess of Warwick Mine) was kept separate from that of the third voyage as indicated by later production of gold and silver from each of the two voyages. Apparently the same precautions were taken against theft of ore at Dartford as at Bristol and the Tower because, in December, Michael Lok was reimbursed 14s 6d for "5 stocklocks and 6 padlocks for the howses".[50] Richard Rawson, proprietor of the Bull Inn in Dartford, was given the keys.[51] In December 1578, the ore still had value!

Pilot-plant production

Extraction of gold and silver, on a pilot plant scale, began in November 1578. It involved the same principles as the proofs, namely fire assaying and nitric acid and/or sulphide parting. Ore was processed in stages, each complete extraction lasting several days, utilizing large crucibles. Amalgamation had been used by the Spaniards in the Americas, but the process was kept a secret; cyanidation was not developed until the nineteenth century.

Additives were purchased from several sources: lead and litharge (PbO) mainly from merchants in London, galena and 'copslighe' (rich copper ore stamped fine) from the Lake District, 'marquasite' (pyrite?) from Newcastle, and 'copshredds' (a copper ore?) from Leith. The accounts suggest that 'antimonye' (stibnite) from Cornwall may have been used in a small way.[52]

Humphrey Cole, a temporary employee of the mint, well known instrument maker and inventor of mechanical appliances, was sent by order of Privy Council[53] to investigate sources of 'marquasite' for furnace additive. He settled on Newcastle, where 'marquasite' was available from coal beds ('sea coal'). The mineral was loaded into the

Bark Rowe in September and November 1578 and shipped to Dartford.[54] This occupation inevitably earned him the sobriquet Humphrey Sea Cole.

Robert Denham was also under Privy Council direction.[55] His 'copslighe' and galena came from Caldbeck, Cumbria and was purchased from the Mines Royal through Daniel Hechstetter. Two shipments were sent southward in December, packed onto 20 horses, each horse carrying two hundredweight. The Company had to pay extra for "carriers and their horses tarrying" because, on arrival in London, no-one was there to receive them.[56]

Information concerning the five pilot plant tests is summarized in Table 5, Nos 7 to 11. Shutz had originally requested "Newcastle sea cole marquasite" but when the first extraction in the great works (No. 7) failed, the additive was blamed as being "but a boddye of sulfer". After the second extraction (No. 8), using additive from Caldbeck, "the Ewre brought home by Captain Furbusher grewe into great discredit".[57] Following two other disappointing trials, on February 27, 1579, a fifth and final extraction (No. 11) was made, which gave the highest values of all the Dartford tests. After this, Lok reported "The Companye them selves have not proceded anything at all in the works at dartforde, but all lyethe still dead".[58] After a mere 15 weeks and extractions from 17 tons of ore, with argentiferous additive, the Company had managed to produce a few ingots of silver and gold (in fact almost completely silver), weighing in all 354 ounces or an average of 21 ounces to the ton.[59]

Later developments

The tests at Dartford had not been a success and the results were even less encouraging than those at London. Shutz blamed the equipment and additive. Consequently, it was decided to return to the London furnaces for one last proof using "the same order w[hi]ch they had done the yeare before" and thus attempt to repeat the higher values. On March 22, 1579, a new slate of officers convened at Sir William Winter's house to observe the pouring of precious metals (No. 12, Table 5). In spite of pains taken to avoid any loss of gold and silver in the slag, there was hardly enough metal to compensate for expenses of mining, transportation and processing; it registered but £15 a ton.[60]

Now followed a period of inactivity, during which Shutz seized the opportunity to propose purchasing the ore himself. He would buy it at 20 marks (£10) the ton. This offer was flatly rejected by the commissioners, mainly through the vigorous opposition of Frobisher, who still had faith in the merit of the ore.[61]

As it was, the operation of the Dartford works had been far from harmonious. Frobisher had been involved in heated arguments with members of the Fenton club: Michael Lok, Jonas Shutz, Edward Castelin, Richard Rawson, and Edward Fenton himself. The barrenness of the ore was becoming increasingly evident and, with this awareness, tempers became short. To make matters worse, Frobisher suspected the mill men and their associates of dragging their heels. William Hanke, Frobisher's cousin and servant, whom Lok did not trust, was the new keeper of the keys. On April 16, Edward Sellman, 'merchant' in the second voyage, 'registrar' in the third, and 'one of the Commissioners' servants', was made master of the works, in hope of restoring order and good will.[62] But the premises remained deserted and closed for another year.

In a mysterious report of March 21, 1580(n.s.)[63], for which we know neither requisitor nor purpose, Daniel Hechstetter and George Needham summarized the state of the Dartford smelter and its ability to melt additives. They tested large samples of copper (2.4 tons) and lead ores (0.3 tons) from the Lake District, with prior roasting in 'Mr Popes soulpher potts' at Queensboro. Possibly this was an attempt to test the efficiency of the Dartford smelter, perhaps with an aim of foregoing smelting at Keswick, where there had been technical difficulties, and using the Dartford smelter to produce metals for the market in southern England. They met with some success in the trials of lead ore from Caldbeck but their trial of Cumbrian copper ore was a dismal failure. Their conclusions concerning the Dartford furnaces are especially pertinent:

> in smelting of the said [copper] yeures [ores] we found such want in the buildinge of the furnace and in the disorderly placinge of the bellowes that we coulde not by any meanes possible p[er]fectly smelte downe the sayd yewrs. If the bellowes and blaste had been good we might have gotten iij qtrs [3/4] or at least Di[th] [1/2] of the copper w[hi]ch we doe nowe loose.

This is the last we hear of these negotiations.

However, Michael Lok did not relax his attempts to wrestle the ore and Dartford works from the hands of the Company. A new offer was made in February 1581: Lok would pay the Company 5 pounds for every ton of northwest ore processed, until all '1200 tons' [1300 tons?] were exhausted. In return Lok would have full use of the 'great works', the Company would grant a *quietus est* for 3 years and pay Lok 10s a day for his labours, as well as provide a bookkeeper, 'additament' and privileges to mine ores in England.[64] By this time, the full burden of the Company's debt had been placed on Lok's shoulders, and this offer was a means of squaring accounts. But the proposal fell on deaf ears and Lok went to debtors' prison.[65]

In July 1583, a final attempt was made to reinstate the value of the ore by William Williams, warden of the mint at the Tower.[66] He made two proofs, each from a sample of 1 cwt (Table 5, Nos 13 and 14) using 2 and 4 cwt lead. He even included "adytament

... to trye the uttermost [to discover] what was in it". The certificate is complete with two little beads of silver still adhering to the wax in the margin, in size proportional to the lead used. No information could be found suggesting who made this order or who paid for the service.

Assays and Metallurgy

Early in the sixteenth century, Germany was a major centre of metallurgy but by the close of the century German technology had spread throughout Europe. The expertise included not only the metallurgy of iron, copper and zinc, but also that of gold and silver. The art of assaying for precious metals was particularly well developed and resembled fire assaying today. Even Burchard Kranich's 'parting' of gold from silver with 'antimonye' (stibnite) was used at the Dresden mint up to mid-19th century (Percy 1880: 367-73).

Fortunately, through the writings of Agricola (Hoover & Hoover, 1950) and Ercker (Sisco & Smith, 1951) we have a good grasp of German metallurgy in the 16th century. We can even establish some of the details and eliminate ambiguities, through examination of expense accounts in the Exchequer papers. For example, we know that the furnaces were built with cemented bricks and fired with coal. They were apparently square and reinforced with iron bars. However, details of the crucible charge were commonly disguised in such nebulous terms as 'additaments' and 'things necessarye for the meltinge'. It is not until we closely examine the expense accounts following the third voyage that we are able to trace the procedure, with confidence, from raw ore to isolated gold and silver. A resumé of the technology involved in the old assays is given in Appendix 1.

The inflated results of these early assays cannot be explained with certainty, given the evidence at hand. The assayers, and even Frobisher and Lok, have been accused of dishonesty but it would seem that neither assayer nor administrator had much to gain from such a major fabrication. It has also been suggested that the mistake lay in incompetence or at least inability to cope with the problems inherent in large assays of an ore of this complexity (see McDermott 1984: 116-117, 129). This is another possibility that cannot be summarily dismissed. Certainly the furnaces were poorly constructed and not up to the task and some of the parting methods were ill-chosen (see Appendix 1). Hogarth & Loop (1986) have suggested that accidental addition of argentiferous lead may have, at least in part, been responsible for the exaggerated values of these early assays.

With regard to the last possibility, it is instructive to examine, in detail, some of the larger metallurgical tests. At Dartford, additives included lead and copper ores imported from Keswick, Lake District. For example, the large test of February 17, 1579 (10 cwt; No. 11, Table 5) consumed at least 13 cwt of Caldbeck galena[67]. The Company of Mines Royal had extracted annually during the preceding decade, 400 ounces of silver, mostly from Caldbeck lead ore. This ore averaged 25 to 50 ounces of silver the ton[68]. Such an additive to the furnace charge, adequately accounts for the 354 ounces produced at Dartford in 1578-9, but not the high-grade assays, as discussed in the next chapter.

Notes

1. M. Lok, memorandum, 1581[?], *BL*, Lansdowne MS 100/1.

2. Data of the first great proof (Table 5, No. 1) are derived from Michael Lok's statement (*PRO*, SP 12/122/62). The assay quoted in Chancery Miscellanea (*PRO*, C 47/34/6) represents a 4 cwt sample, which gave 0.5 ounces gold and 8 ounces silver per ton.

3. This quotation, as well as many details pertaining to the assays of J. Shutz, are given by M. Lok, Feb. 1578. *PRO*, SP 12/122/62; *SM* 2: 123-8.

The phrase "gold remayning in the slags" presumably implies incomplete fusion although it is hard to understand how Shutz was able to place a quantitative value on gold in the slag.

4. M. Lok to Sir F. Walsingham, Jan. 19, 1578. *PRO*, SP 12/122/9; *SM* 2: 120.

5. The storehouse of Sir William Winter is quoted in *BL*, Lansdowne MS 100/1, ff. 4R, 4V, 13R, as well as *PRO*, SP 12/122/62; *SM* 2: 123-8, and elsewhere. The hypothesis of its location, just north of the church of All Hallows Barking, is confirmed by assessment records held in the church (Redstone *et al.* 1929: 19-20). The statement that the ore was kept in "the Quenes storehous" (*PRO*, E164/35, f. 149), is misleading, as is apparent from other data within the passage (*ibid.*, ff. 149-153). Its site on "St Katheryns hyll" (*PRO*, SP12/115/35; *SM* 2: 109-110) appears to be in error.

6. Both sets of items are taken from the audited accounts of M. Lok, 1577. *PRO*, E 164/35, f. 149. Itemized accounts are also available for the two succeeding 'great proofs'.

7. The total weight of ore assayed (4 cwt) and the amount of gold and silver recorded (Fig. 15; *PRO*, C47/34/6) suggest *partial proceeds* from *all three* great proofs of Jonas Shutz (cf. Table 5, anal. 1, 2, 6).

Allen's bark returned with 166 tons of ore in the third voyage. As late as August 12, 1580 he was owed £392 5s 2d for his cargo (*PRO*, E 164/36, f. 331).

8. In Lok's accounts there is reference to ore stamped at Muscovy House but elsewhere it appears that the assays themselves were made on Sir William Winter's property (*PRO*, E 164/35,

ff. 149-238). There is no firm evidence for furnace construction at Muscovy House although the 'third great proof' was certainly registered there March 8, 1578, three days after completion. (*PRO*, SP12/123/5; *SM* **2**: 131-2).

9. Data for this paragraph are taken from M. Lok, 1578, [The Account of] *small profes and sayes made at London of the mynerall Ewer brought home by M[aste]r furbusher in this second voyage, PRO*, E 164/35, ff. 149-54; M Lok, 1578, *The causes and foundacyon of the iij voyages, PRO*, SP 12/149/42 (iii); J. Shutz, 1578, *Thaccompt taken at muskovie house the viij of march 1577, PRO*, SP 12/123/5; *SM* **2** 131-2, copied in *PRO*, SP 12/149/42 (ii).

Gold from one of the 'great proofs' (probably the first) is preserved as a large flattened bead on the assay sheet (*PRO*, SP 12/119/9; Fig. 16, this monograph). Its rather pale yellow colour suggests that removal of silver with HNO_3 'parting' was incomplete.

10. This, and further biographical information, may be found in Donald (1950, 1951). Kranich was also known as Dr Burcot[t], especially in his medical career. He spelled his name Kranrich. In this publication he is referred to as Burchard Kranich, consistent with M. B. Donald's biographical sketch.

11. M. Lok, memorandum, Feb. 1578. *PRO*, SP 12/122/62; *SM* **2**: 127.

12. B. Kranich to Sir F. Walsingham, Feb. 27, 1578. *PRO*, SP 12/122/61; *SM* **2**: 130.

13. Patent to J. Shutz, Jan. 11, 1578, Calendar of Patent Rolls, Elizabeth I, 7 (3668). Details of the patent to R. Denham and B. Kranich are not given in the calendared rolls, but summaries are found in *BL*, Landsdowne MS 100/1, ff. 4V, 5R, (1581 ?). Payment of patent fees out of Company funds, to all three, are itemized in Lok's expense account *PRO*, E 164/35, ff. 151-2 (1578).

14. The quotations are from J. Shutz, Feb. 1578, *The dooings of M[aste]r Burcott in the newe mynes of Gold, PRO*, SP 12/122/62; *SM* **2**: 127-8.

15. B. Kranich, Dec. 9, 1577, *Mr. Dr. Burcot's Articles and Conditions, PRO*, SP 12/119/12 (i, ii); *SM* **2**: 145, 146; M. Lok, June 16, 1581, *The causes and foundacyon of the iij voyages, PRO*, SP 12/149/42 (iii).

16. J. Shutz, Feb. 1578, *The dooings of M[aste]r Burcott in the newe mynes of Gold, PRO*, SP 12/122/62; *SM* **2**: 127-8.

17. B. Kranich to Sir F. Walsingham, Feb. 27, 1578. *PRO*, SP 12/122/61; *SM* **2**: 130.

18. *APC* **10**, Mar. 6, 1578. p.177.

19. E. Fenton to Privy Council, Jan. 2, 1579. *PRO*, SP 12/129/2; *SM* **2**: 147-9. M. Lok, *PRO*, E164/35, f. 255. Fenton visited Cornwall at Christmas-time 1578; his guide was William Paynter.

20. E. Fenton to Privy Council, Jan. 13, 1579. *PRO*, SP 12/129/6. The probable location is given as Wheal Leigh, Pillaton, Cornwall, re-opened with a small production of stibnite from 1819 to 1822. *BM*, MS Sir A. Russell 1949-50. *An account of the Antimony Mines in Great Britain and Ireland*, ff. 24-25.

21. M. Lok to Sir F. Walsingham, Dec. 13, 1577. *PRO*, SP 12/119/12; *SM* **2**: 145.

22. M. Lok, Account of 1578. *PRO*, E 164/35, f. 151.

23. M. Lok, Accounts of Nov. 16, 1578; Dec. 14, 1578; Jan. 3, 1579; Mar. 13, 1579. *PRO*, E 164/36, ff. 183, 184, 186; *PRO*, E 164/35, f. 258. Three of the German metallurgists were Christopher Ryer (melter), Jacob Walter and Andrew Bucker.

24. M. Lok, Accounts of 1578. *PRO*, E 164/35, ff. 150, 152, 153.

25. J. Shutz, memorandum, *c.* Nov. 25, 1577. *PRO*, SP 12/118/42; *SM* **2**: 141. This was a preliminary estimate, made before the Dartford site was selected.

26. M. Lok to Sir F. Walsingham, Dec. 13, 1577. *PRO*, SP 12/119/12; *SM* **2**: 144.

27. Petition of W. Vaughan to Privy Council, Dec. 20, 1577. *PRO*, SP 12/119/14.

28. Account of M. Lok for charges after the second voyage, (1577). *PRO*, E 164/35, f. 151. "Paid to Nicholas Chancelor for his charges ... to the myles at dartford ... li 1.0.0"

29. T. Fludd to Lord Burghley, Jan. 7, 1578. *PRO*, SP 12/122/4.

30. Accounts of John Hales and Thomas Kennyon for expenses at Dartford, 1578-9. *PRO*, E 164/36, ff. 253, 256.

31. This paragraph refers to Lease in Reversion, William Vaughan, 1578, *PRO*,

E 310/ $\frac{40}{5}$ /5; partly English, partly Latin.

Other land papers, in Latin, re William Vaughan and Bignores are:
1. Books of the Court of Augmentation, Aug 4, 1547, *PRO*, E315/217/143; *Cal. Letters and Papers, Foreign and Domestic*, Henry VIII **21** (2): 439.
2. Patent Roll, Oct. 29, 1557, *PRO*, C66/923/m.7; *Cal. Patent Rolls, Philip and Mary* **4**: 118.
3. Patent Roll, Nov. 15, 1570, *PRO* /1064/m.23; *Cal. Patent Rolls, Eliz. I* **5**, No. 317: 56.
4. Patent Roll, May 2, 1578, *PRO*/1169/m.25-6; *Cal. Patent Rolls, Eliz. I* **7**, No 3269: 7 8.

32. *APC* **10**, Jan. 8, 1578: 135-6.

33. J. Shutz, memorandum, [Feb. 1578]. *PRO*, SP 12/122/62; SM **2**: 126. Accounts of M. Lok for expenses at Dartford, payment of April 5, 1578. *PRO*, E 164/36, f. 177.

34. M. Lok to Sir F. Walsingham, Jan. 19, 1578. *PRO*, SP 12/122/9; *SM* **2**: 119-20. The estimated cost of materials and labour was £900.

35. Accounts of M. Lok for expenses at Dartford, payments Feb. 11, 1578 *et seq.*, *PRO*, E 164/36, ff. 177 *et seq.*

36. *APC* **10**, July 3, 1578, pp. 271-2. In order to meet expenses at Dartford, in 1578 venturers were assessed 135% on their original investment (later re-assessed 85% on both stock and assessment). The Queen's share in the first assessment came to £1350, of which £200 were defrayed for her buildings and property (*PRO*, E 164/35, f. 152).

37. Account of E. Castelin for debts owing at Dartford, 1579. *PRO*, E 164/36, ff. 201-99.

38. Audit of Thomas Neale, Jan. 1579. *PRO*, E 164/35, f. 72.

39. M. Lok to Sir F. Walsingham, Jan. 19, 1578. *PRO*, SP 12/122/9; *SM* **2**: 120.

40. Reference is made to Kranich's drawings ('platts' or 'patterns') in the following manuscripts concerned with the Frobisher enterprise: *PRO*, SP 12/122/53 (Feb. 21, 1578); *PRO*, SP 12/122/44 (Feb. 19, 1578); *PRO*, SP 12/122/63 (n.d.); *BL*, Landsdowne MS 100/1 (*c*. 1581). The diagram is thought to be the second version, prepared Feb. 20, 1578, attached to SP 12/122/53, and sent to Walsingham Feb 21, 1578. The drawing is now catalogued as *PRO*, MPF 305.

Some historians attribute this drawing to lead smelting in Derbyshire because of correspondence of the plan with contemporary descriptions of the works of *c*. 1555 and because a 'model' of a smelter at Beauchief, Derbyshire was described to Lord Burleigh in 1582 (Donald 1961: 166, 169; Kirkham 1969).

41. J. Shutz, memorandum, Feb. 1578. *PRO*, SP 12/122/62; *SM* **2**: 127.

42. Accounts of M. Lok for expenses at Dartford, payments of May 14, Aug. 4, Sept. 14, Sept. 30, Nov. 5, Nov. 28, 1578. *PRO*, E 164/36, ff. 177, 180-184.

43. Summary account of M. Lok, May 6, 1581. *PRO*, E 164/36, f. 327.

44. *APC* **10**: 373-4, July 3, 1578.

45. Account of M. Lok for expenses at Dartford, payments of July 22, 1578 and Jan. 3, 1579. *PRO*, E 164/36, ff. 179, 186.

46. M. Lok to Sir F. Walsingham, Dec. 11, 1578. *PRO*, SP 12/127/16; *SM* **2**: 172-3. On December 11, the consignment of stibnite had not been officially requested.

47. Account of M. Lok for expenses at Dartford, payments of Nov. 4-28, 1578. *PRO*, E 164/36, ff. 183-4.

48. Account of E. Castelin for debts owing at Dartford, 1579. *PRO*, E 164/36, f. 210.

49. *ibid.*, f. 211.

50. Account of M. Lok for expenses at Dartford, payment of Dec. 30, 1578. *PRO*, E 164/36, f. 185.

51. M. Lok, memorandum, *c* 1581. *BL*, Lansdowne MS 100/1, f. 10V.

52. Account of T. Allen for expenses after the third voyage, 1579. *PRO*, E 164/35, f. 261. "Paid ... to chandler the 2 of may for charges in sending downe the Adittament to dartforde that came from m[aste]r fenton out of the west cun[trie] 7s". Account of M. Lok for expenses at Dartford, 1578-1579. *PRO*, E 164/36, f. 185. "The 14 december 1578 paid to Jonas for Addittaments for plymouth 10s".

53. *APC* 10, Feb. 3, 1578. p. 156; *APC* 10: 272, July 3, 1578.

54. Account of M. Lok for expenses at Dartford, 1578-9. *PRO*, E 164/36, ff. 181-3. "The 12 september 1578 paid [Michael Lok] for going abourd the bark Rowe at the dischardging of the m[ar]quesite from newe Castell 4d; the 24 september paid Mark Rowe pursar of the bark Rowe for fraight of 14 tonnes of Cole m[ar]quesite ... 5 li; the 6 november 1578 paid to [H]umphry Cole for chardgs of m[ar]quesyt w[hi]ch came from newe Castell 4 li; the 6 november 1578 paid for lighterage of 16 tonne Se[a]cole m[ar]quesyte to dartford, paid to John Kymber 1 li 4 sh".

55. *APC* 10: 398, Nov. 22, 1578.

56. Account of M. Lok for expenses at Dartford, payments of Dec. 30, 31 1578. *PRO*, E 164/36, f. 185. "The 30 december paid to Anthony Frekleton and Robert Lockey carriars of kendall for carriag[e] of xj hor[se]loads of Ewre wh[er]of 6 loads carriag[e] from keswick to kendall and so to London at xxj s. viij d the load ij c waight the horse load ... li 6.10s for thother 5 Load at xx s aload from kendall to London ij c waight eche ... li 5. for chardgs of the said carriars and their horses tarrying 6 daies for their money in London ... li 1."

57. M. Lok, memorandum, *c* 1581. *BL*, Lansdowne MS 100/1, f. 10R.

58. *Ibid.*, f. 14R.

59. At Dartford, three batches of alloyed silver and gold are recorded in the Exchequer Papers (*PRO*, E 164/36, f. 333): i, 'certaines pecs' melted and refined by Jonas Shutz, from part of a 16-ton lot taken from the *Judith* - the product was an alloy, weighing 120 ounces and containing 98.8% silver and 1.2% gold; ii, a silver-gold alloy, refined by Robert Denham, the result of a previous extraction made by Jonas Shutz from the remainder of the 16 - ton lot (The final product weighed 210 ounces); iii, a silver-gold alloy, produced by Robert Denham from a separate 1-ton

lot of ore and weighing 24 ounces. Individual gold and silver separations ('partings' of the alloys) were recorded for batch i only.

60. M. Lok, memorandum, *c*. 1581. *BL*, Lansdowne MS 100/1, f. 13R.

61. *ibid*., ff. 13R-13V. Shutz, Denham and the German workmen offered to buy all the ore at 20 marks the ton, saving 150 tons in a separate offer, which Lok would work himself. The two offers were submitted to Burghley and Walsingham April 18, 1579, who relayed them to the Council. On the advice of Frobisher, who believed the ore was worth at least £40 the ton, these offers were rejected.

62. *ibid*., f. 13R.

63. Declarations of D. Hechstetter and G. Needham, March and April 1580. *PRO*, E 164/36, ff. 307-13.

64. The offer of M. Lok for purchase of the northwest ore, Feb. 1581. *BL*, Lansdowne MS 30, ff. 10-12; *SM* **2**: 204-206.

65. A well documented biography of Michael Lok is presented by McDermott (1984).

66. Assay certificate of W. Williams, July 28, 1583. *PRO*, SP 12/161/41.

67. M. Lok, Accounts, Nov. 1578 to Jan. 3, 1579. *Charges paid at London by Michael Lok for dyvers things for the byldinges at dartford and for the Workes there*, PRO, E 164/36, ff. 182-6.

The 10 - cwt test (No. 11, Table 5) was made from ore taken from the *Judith* (third voyage) and entirely derived from Countess of Sussex Mine.

68. Source: Daniel Hechstetter and, his son, Emanuel in Hammersley (1988: 171-2). Silver produced in the Lake District by the Company of Mines Royal from 1570 to 1580, inclusive, was 4344 ounces. The grade, for all but the leanest lead ores, ranged between 2.5 and 5 lotts [loths, *OED*] per hundredweight.

CHAPTER 5

Ores and Mines

At the present time, Frobisher's ore is known from three areas: southeastern Baffin Island (Canada), the Borough of Dartford (England), and Smerwick Harbour (Ireland). Its occurrence and description have received passing notice only. Although this chapter will summarize existing data, it is not our intention to present a detailed discussion of the mineralogy and petrogenesis of these rocks; the results will be submitted for publication elsewhere. Rather, we will include descriptive data from on-site and laboratory investigations and put forward a few ideas on the nature and origin of Frobisher's 'black ores'.

Rediscovery of 'black ore' on Kodlunarn and Baffin Islands, 1861-1985

As a by-product of the search for remnants from the ill-fated Franklin voyage, Charles Francis Hall discovered Frobisher's landfalls and located the site of a mining operation on Kodlunarn Island (Countess of Warwick Mine, Fig. 21). He identified an inland trench, the Reservoir Trench, as a possible mine site and discovered coal heaps that had been left on Kodlunarn and nearby islands (Hall 1864 2: 150-6). A trench on the north shore of Kodlunarn Island, the Ship's Trench, was described as an excavation used to repair and build ships - i.e. it served as a makeshift drydock.

The next scientific investigation was by Sharat Roy, the geologist with the Rawson-MacMillan interdisciplinary expedition of 1927. Roy (1937) noted a stockpile of black rock near the Ship's Trench but, as he was unable to locate similar rock *in situ* nearby, concluded the fragments were "apparently foreign to Kodlunarn" and had been placed "on the edge of the island, where they could be readily loaded into the ships' holds". He collected and described a specimen of black 'amphibolite' from this storage heap and dark green 'pyroxenite' from "an outcrop about two hundred yards" from Kodlunarn Island.[1] Roy believed that Frobisher mined various types of dark igneous and metamorphic rock and that it was a bronzy biotite (rather than pyrite) that attracted the attention of miners and assayers.[2]

As part of a program of reconnaissance mapping (1:500,000) of the Arctic islands by the Geological Survey of Canada, R.G. Blackadar visited Kodlunarn Island in 1964 and 1965. He found bedrock on the northeast side of Frobisher Bay to be a light-coloured gneiss [3] (Blackadar 1967a) but, in a description of Kodlunarn Island, Blackadar (1967b) stated that the trenches penetrated layers of 'amphibolite' containing small flakes of biotite. He noted that the two trenches were aligned, suggesting that Frobisher mined amphibolite from a single horizon in both trenches.

In 1974, Kodlunarn was visited by W.A. Kenyon and associates, representing the Royal Ontario Museum. The investigations were mainly archaeological, but Kenyon returned with 50 kg of rock. He also illustrated the tell-tale pick scars left by Elizabethan miners and described coal deposits at Victoria Bay, Baffin Island, previously noted by Hall (Kenyon 1975a). In addition, he pinpointed several possible Frobisher quarries on mainland Baffin.

Visits to Kodlunarn Island were also made by Hogarth and Gibbins for the Department of Indian Affairs and Northern Development of Canada in 1975, 1983 and 1985. The following publications have resulted from this research: geology of Kodlunarn Island (Hogarth & Gibbins 1984; Hogarth *et al.* 1985), preliminary classification of Frobisher ores (Hogarth 1985), precious metals in the ores (Hogarth & Loop 1986). The most important conclusions were:

1.   Frobisher mined both trenches at Kodlunarn Island, but the Ship's Trench was essentially finished in 1577. In 1578, mining was mainly centred on the Reservoir Trench.
2.   Stockpiles of ore near the Reservoir Trench were derived locally, whereas pieces of ore near the Ship's Trench were derived locally, and from other mines.
3.   Fenton's Fortune of Kenyon (1975b: 175) is not Fenton's Fortune of Frobisher, which was on the opposite side of Countess of Warwick Sound. The former is a doubtful mine and probably represents a sea cave.
4.   The principal mineral component of Frobisher's 'black ore' was hornblende. A feldspar (calcic plagioclase), pyroxenes (diopside and enstatite), mica (biotite) and spinel (mainly hercynite) were present in lesser amounts.
5.   Frobisher tested 'black ore' at Kodlunarn Island and Little Hall's Island by oxidizing fine-grained greenish mica and enstatite to golden minerals by ignition in air at red heat. The rock was initially roasted in the ship's furnace.
6.   Frobisher's largest mine, the Countess of Sussex Mine, was tentatively pinpointed on Baffin Island, 9 km west-northwest of Kodlunarn Island.
7.   Except for a very few early proofs, the Elizabethan assays of Frobisher's ore suggest silver, not gold mines (both in metal content and in contemporary sterling value).
8.   The Elizabethan valuation of precious-metal content was grossly inflated, possibly due, at least in the later assays, to inadvertent addition of contaminant to the furnace charge. The high values reported in the early assays may have been due to faulty analytical procedures.

These conclusions will be expanded or modified below.

'Black ore' at Dartford

The Frobisher rock at Dartford has been known for many years. Fortunately, specimens that came to light in the 1920s and 30s were spotted by amateur archaeologists. In 1921, when making a cable trench from the pattern shop of Hall's Engineering works on Priory Road, the foundations of the old wall of the Manor House were breached, exposing large blocks of black, hornblende-rich rock. Samples were sent to Cambridge where they were examined by Professor V.C. Illing who reported augite, olivine, plagioclase and biotite associated with hornblende. A highly weathered, black Dartford specimen was sent to the Mineralogical Laboratory at Cambridge in 1926. Dr A. Hutchinson made a cursory examination but its friable nature rendered it unsuitable for detailed mineralogical work. He requested fresh material but, at the same time, concluded "there can be little doubt that the specimen [represented] part of what Sir Martin brought" (unpublished notes in the Dartford Borough Museum). The Cambridge geologists knew that Frobisher's rock had been stored in Dartford and this was a logical conclusion. Further specimens were, in 1930, recovered from a drainage ditch near the Thames estuary and sent to Cambridge, where Dr R.E. Priestley identified them as 'micaceous amphibolite' (*SM* 2: 249). A large black specimen, acquired later, was examined by Dr C.E. Tilley and found to be micaceous 'hornblende pyroxenite' (Roy 1937: 34). Two thin sections, now in the Harker collection at Cambridge, perhaps from specimens collected about the same time, are also 'hornblende pyroxenite'.[4] Dr C. Ritchie (1964) published a photograph showing a block of 'peridotite' in the old wall surrounding the Dartford Manor House. Obviously the Dartford rock comprises an assortment of rock types.

A surprisingly large quantity of the 'black ore' has been preserved in the borough. Some is incorporated in a 120-m section of Tudor boundary wall fronting the Priory Road. This boundary wall was constructed some time after the spring of 1579 to reinforce the western perimeter of what was formerly King Henry VIII's Manor House. Other sections of walling around the site are mediaeval, 19th and 20th century. Only a small fragment of the Manor House survives today, representing part of the western gatehouse.

More than 530 pieces of 'black ore' are visible on the exterior face of the wall fronting Priory Road. Blocks of ore in excess of 30 x 30 cm are not uncommon (Fig. 20). A small quantity of 'black ore' was also used in patching some of the 19th and 20th century reconstructions of the wall.

In 1976-7 and again 1982, the Dartford District Archaeological Group uncovered specimens of 'black ore' in excavations at the site of the Manor House (Fig. 17). Those of 1976-7 (sample series E23) were found in a small area about 300 m north of the surviving gatehouse and near the west wall. Specimens found in 1982 (sample series E44) were retrieved from a pit nearby, which also contained Tudor floor tiles.

Figure 20: The western wall of the former Dartford Priory, Kent, showing large blocks of Frobisher's 'black ore' used in its reconstruction.

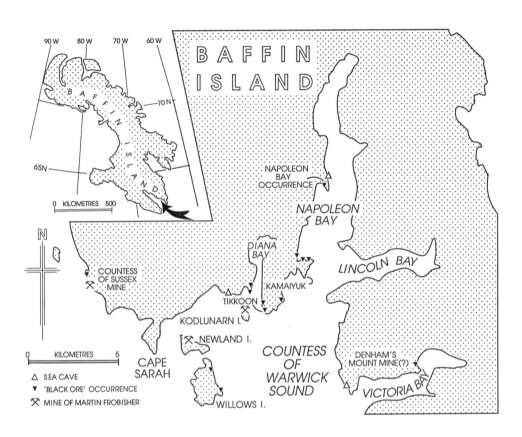

Figure 21: Frobisher mines and sample sites, Countess of Warwick Sound and vicinity, southeast Baffin Island. Geographic names are those of present usage.

'Black ore' has been excavated from other localities in Dartford, most notably the site of the former Bull and George Inn, Dartford High Street (sample series E43). Here specimens of ore were retrieved from primary fill in a 16th century cess pit. A small quantity of 'black ore' was also retrieved in 1988 from a site at Home Orchard, Bullace Lane, close to Dartford High Street (sample series E47). Additional blocks of 'black ore' were discovered in an excavation in a private garden, King Edward Avenue, Dartford, in 1984, approximately 250 m north of the Dartford Road (sample AC). The distribution of 'black ore' in Dartford supports the theory that the rock became available for general building purposes after smelting operations ceased.

'Black ore' at Smerwick, southwest Ireland

The solid or native rock at Dún-an-Óir (Chapter 3) comprises a succession of sandstone, conglomerate and tuff (Horne 1974, 1976; Todd *et al.* 1988), with strata separated by fault and unconformity, and ranging in age from Devonian (*c.* 400 million years) to Silurian (*c.* 450 m.y.). Cradled between cliffs of these Palaeozoic rocks are coves with sandy beaches, exposed at low-to-medium tide and fringed by cobble strands. The cobbles are mainly derived from the local rock but a few are quite distinct and bear no resemblance to the *in situ* exposures. Those collected in 1987 were described (Hogarth & Roddick 1989) as: *A*, hornblende with diopside and ilmenite (distinctive trace quantities of vanadium, little chromium), and *B*, hornblende and forsterite (distinctive trace quantities of chromium, little vanadium). Type *B* was also found on Baffin Island and at Dartford. A black 'banded ore', consisting of layers rich in hornblende and forsterite (black) alternating with layers rich in diopside (green), is common in suites from all three regions. These rocks contrasted with type *C*, hornblende with plagioclase from Kodlunarn Island (distinctive trace quantities of nickel but little chromium and vanadium). The *Emanuel* did not load at Kodlunarn (Chapter 3), which explains the absence of *C* ore at Smerwick. Types *A* and *B* do not belong to Ireland or the British Isles, in general.

The Baffin Island area

Southeast Baffin Island is predominantly a terrane of Precambrian gneiss (Blackadar 1967a). There is little sand or gravel cover. Hills commonly rise abruptly 300 to 400 m above the sea, in places as cliffs, so that the rock is well exposed and easily examined in three dimensions. The gneiss surrounding and close to Countess of Warwick Sound, i.e. the area of Figure 21, is normally a hard, fine-grained, foliated, greyish white rock, containing little besides quartz, feldspar and biotite. In such exposures, only dykes of coarse-grained pink granite (pegmatites) interrupt this monotonous regularity. However

Table 7: *Frobisher's mines, listed in order of discovery*

Mine	Approximate location	Discovered	Worked	Type of ore	Remarks
Little Hall's I.	N side, mouth of Frobisher Bay	By R. Garrand Aug. 10 1576		'Black'	Initial discovery, July 1577. Attempts to locate ore *in situ* unsuccessful.
Jonas Mount	15 km NNE of Kodlunarn I.	By G. Bona & J. Shutz 1577		'Yellow' & 'red'	High Au assays. Sand and small pieces only. Attempts to locate more ore unsuccessful.
Leicester I.	Beare Sound	July 26 1577	July 26 1577	'Black'	20 tons mined 1577, ready for loading 1578.
Countess of Warwick	Kodlunarn I.	July 29 1577	July 29-Aug. 21 1577; Aug. 1-9 1578	'Black'	Headquarters of mining operations 1578. Assay furnace site 1578.
Winter's Furnace	4 km SSW of Kodlunarn I.	By E. Fenton July 21 1578	July 21-c.25 1578	'Black'	First assay furnace of 1578.
Fenton's Fortune	15 km SSE of Kodlunarn I.	By E. Fenton July 29 1578	Aug. 9-15 1578	'Black' some 'white'	R. Philpott in charge. Small tonnage available.
Sussex I.	Beare Sound	Aug. 7 1578	Aug. 9 - Sept. 1 1578	'Black'	Principal mine in Beare Sound.
Best's Blessing	S-side, mouth of Frobisher Bay	By G. Best Aug. 9 1578	Aug. 10-25	'Black'	Large tonnage available.
Countess of Sussex	9 km WNW of Kodlunarn I.	By T. Morris & R. Davis	Aug. 11-25 1578	'Black' 'red' & mixed	Ore considered high grade.
Denham's Mount	9 km ESE of Kodlunarn	By A. Diar	Aug. 19-29 1578	Four types	Furnaces installed nearby.

layers, containing minor hornblende (imparting a buff colour), enstatite (rusty), sillimanite (yellow) or garnet (pink), are not rare. Layers of dark-green-to-black gneiss, made up with hornblende, diopside, enstatite, biotite and plagioclase, in varying proportions, are locally common, and at Countess of Sussex Mine, Tikkoon and Napoleon Bay, they are closely associated with 'black ore'. The whole assemblage normally strikes north-northwest and dips westerly at moderate to steep angles (45-60°).

Three of Frobisher's mines of 'black ore' can now (1993) be identified (*viz.* Countess of Warwick, Countess of Sussex and Winter's Furnace). Information concerning the original discovery and operation of these and other Frobisher mines is summarized in Table 7. 'Black ore' is most obvious at other occurrences in the region (e.g. Tikkoon Point and Napoleon Bay) and must have been known and tested by Frobisher's men in 1578. Occurrences are located in Figure 21.

Countess of Warwick Mine, Kodlunarn Island

Countess of Warwick Mine on Kodlunarn Island (Frobisher's Countess of Warwick Island) was discovered July 29, 1577[5] and it produced 158 tons of 'black ore' that year. The first mine seems to have been the Ship's Trench, at the north end of the island. This trench was essentially finished in 1577 because its outline, today, corresponds closely to that described by Count Mendoza after the second voyage (Hume 1894: 567-569). Mining during the third voyage, therefore, was probably confined to the interior, or Reservoir Trench (Fig. 22); but this opening was abandoned in favour of other mines, after a week's arduous work. It had produced 65 tons, from tough unyielding rock, that were loaded into three barks (tonnage estimate of E. Sellman in *SM* 2: 69-70). The island then became the centre of administration and assaying for the duration of this expedition (three weeks), after which it was lost to Europeans for nearly 300 years.

On Kodlunarn, 'black ore' can be traced along the eastern part of the island for 200 m (Fig. 23). It crops out below the Ship's Trench and appears to have once been exposed in a trench in the middle (the Reservoir Trench). In 1991, a small test pit (now filled in) was excavated below the Reservoir Trench (Fig. 23). At about 50 cm depth, a deposit of coarse black sand was encountered, composed mainly of hornblende, lesser amounts of other dark minerals, and about 5% feldspar. This sand contained solid kernels of 'black ore' up to first size, four of which were saved for detailed study. The unconsolidated nature of the deposit was, no doubt, caused by shattering due to periodic downward percolation of water in the summer and freeze and thaw in the fall and spring.

The layer of black sand and gravel ran parallel to the trench and tapered at depth. At 50 cm the layer was about 30 cm wide but at a depth of 70 cm it had narrowed to less than 20 cm. The 'black ore' would, therefore, seem to bottom at 2 or 3 metres below the original surface.

Figure 22: The Reservoir Trench, (Kodlunarn Island), 25 m long, excavated in 1578 and supplier of about 65 tons of 'black ore' for the Dartford works. Ore was mainly type *C1*, with lesser amounts of *A6, C2* and *C3* (Table 8).

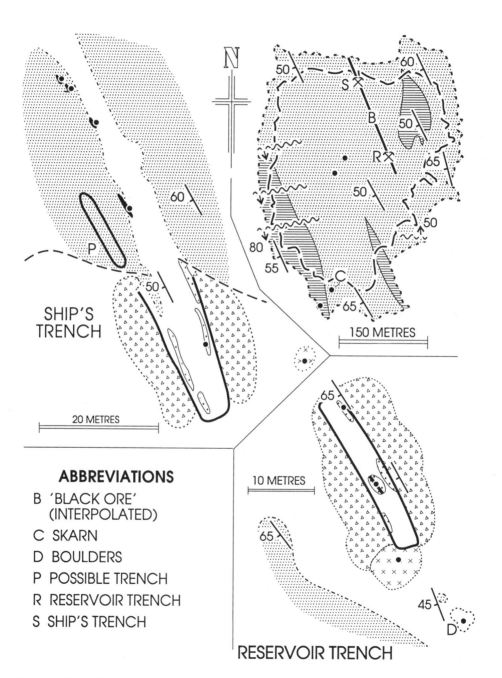

Figure 23: Geology and trenches, Countess of Warwick Mine, Kodlunarn Island.

GEOLOGY OF FIGURES 23, 25 AND 29
PROTEROZOIC

5 GRANITE PEGMATITE

4 'BLACK ORE'. g GARNET - BEARING;
s SPINEL - BEARING

PROTEROZOIC OR ARCHEAN

3 AMPHIBOLE, PYROXENE GNEISS

2 ▭ SILLIMANITE AND/OR GARNETIFEROUS GNEISS

1 ▭ BIOTITE (± GARNET) GNEISS

UNITS 1. 2 AND 3 ARE NOT NECESSARILY IN CHRONOLOGICAL ORDER

SYMBOLS OF FIGURES 23,25 AND 29

• SAMPLE SITE   ⚒ MINE    STOCKPILE

L POSSIBLE LOADING SITE   ◆ TALUS

╱ STRIKE AND DIP (FOLIATION), INCLINED

∿∿ FAULT; INCLINED   ⌒⌒ TRENCH

⌒ GEOLOGICAL CONTACT    SPOIL HEAP

⋯⋯ LIMIT OF OUTCROP   ⌒ TOPOGRAPHIC CONTOUR

- - - - TIDAL LEVEL (FROM AIR PHOTOGRAPH)

⌒⌒ TOP OF SEA SCARP

Figure 24: 'Black ore' layers, below the Ship's Trench, (Kodlunarn Island), that have been stretched and separated into isolated lenses or 'boudins'. The ore (dark) is *C1* in the centre, separated by a rind of *C2* from the surrounding biotite gneiss (light).

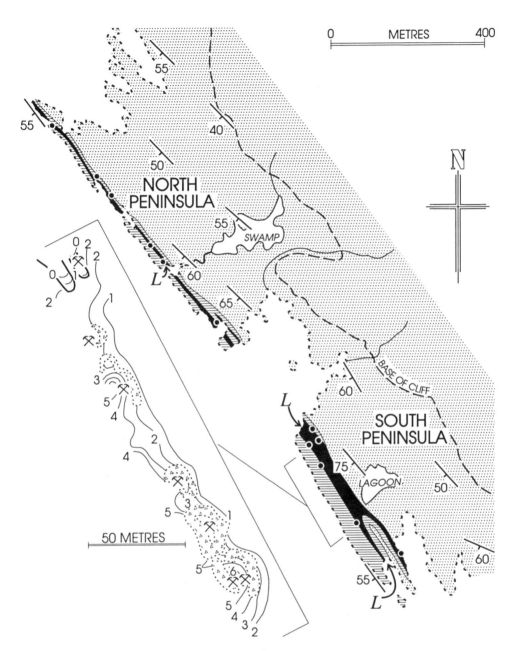

Figure 25: Geology and mine workings, Countess of Sussex Mine, Baffin Island. Contours in the enlargement are in metres above high tide. For legend (geology and symbols) see Figure 23.

Figure 26: Exposure of ore in a Frobisher excavation, on the south peninsula of Countess of Sussex Mine (Baffin Island), worked by Fenton's and Frobisher's companies under Gilbert Yorke, August 1578. This remnant of 'black ore' (type *A1*) may have been abandoned because it was too tough to break. The ridge behind, composed of 'barren' mafite, separates two shallow pits.

Figure 27: Spoil heaps (off ridge in background) on the south peninsula of Countess of Sussex Mine (Baffin Island). Trails of spilled 'black ore' suggest that transportation was by man, cart or wheel barrow, southward (to the left in this photograph), along the flats (below the ridge), to tidewater.

Figure 28: Typical appearance of highly contorted and banded *B1* 'black ore' from the north peninsula, Countess of Sussex Mine, Baffin Island.

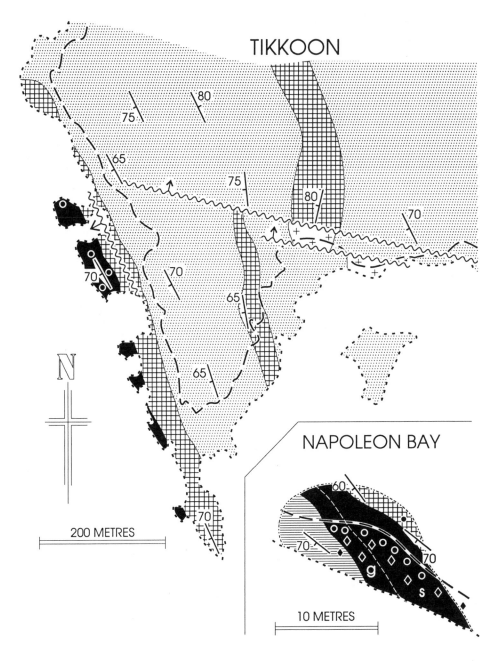

Figure 29: Geology of two possible mines operated by Martin Frobisher in 1578. For legend (geology and symbols) see Figure 23.

The 'orebody', a black interlayer in a rather homogeneous grey gneiss, is extremely hard. This feature reportedly caused the demise of the mine, which "fayled being so hard stone to breke" (E. Sellman in *SM* 2: 70). Add to this, the inconsistent and narrow widths of the payzone, which pinches and swells and, in places, is disconnected into isolated lenses or 'boudins' (Fig. 24). In the intertidal zone, just below the Ship's Trench, a boudin of 'black ore' attains a maximum width of 35 cm. This mine had real problems, regardless of the grade of ore.

The ore itself is composed largely of pitch - black, glittering hornblende, a "blacke stone, much lyke to a seacole in coloure, whiche by the waight seemed to be some kinde of Mettal or mynerall" (G. Best in *SM* 1: 51). The periphery of the ore lenses is micaceous, but the black mica gives way to a black pyroxene (hypersthene) in the interior of the lenses, and both minerals are readily transformed to a golden yellow after roasting at red heat. This chameleon-like property appears to have given the go-ahead to mining in 1577,[6] but whether this test was routinely applied during the mining operation is not known. Loose coal, probably fuel for assay furnaces, is present at several locations. It is known that fire assays were conducted on the island using conventional technology (Hogarth 1993a).

Countess of Sussex Mine, Baffin Island

Countess of Sussex Mine, 9 km (5.5 miles) WNW of Kodlunarn Island and "adjoininge to the Maine", was discovered by Thomas Morris (master of the *Francis* of Foy) and Robert Davis (master of the *Ayde*) on August 10, 1578. From August 15 to August 27 it was worked continuously by about 60 men and during this two-week interval, 455 tons of ore (estimate of Edward Sellman) were loaded into 6 small ships and the galliass *Ayde* (Table 4). It was Frobisher's largest mine. However, pertinent assays were not made here, but on Kodlunarn Island.

The most northerly occurrences were rediscovered by Gibbins and Hogarth during a reconnaissance helicopter survey in 1985. The southern occurrences were rediscovered by Hogarth in 1990 and all occurrences were examined in detail by Ala and Hogarth in 1991. They lie on two adjacent peninsulas (Fig. 25), in the approximate positions given by Sellman (*SM* 2: 65) and Fenton (Kenyon 1981: 65).

The geology of this area is rather simple (Ala 1992). A layer of crumbly black (mafic) gneiss,[7] a well defined marker 6 to 20 m thick, is sandwiched between tan-to-gray biotite gneiss, on the east, and pale tan garnetiferous gneiss, on the west. The assemblage dips steeply to the west. We have traced the mafic gneiss across the two peninsulas for a total of 1450 m, but this includes a 300-m length covered by waters of the bay between them.

Layers of solid 'black ore' (ultramafite) [7] up to 8 m thick (Fig. 26) were restricted to the dark gneiss but were found along its entire length. Compared to Countess of Warwick mine, the occurrences were more easily worked. The 'black ore' was thicker, more persistent, and more easily freed from the fissile wallrocks. However, the continuity was interrupted by minor faults and a number of dykes of coarse-grained granite, which now stand up like walls above the mined ore.

In the north peninsula, the marker horizon was traced for 700 m. In one place, a large hole in dark gneiss extends to tidewater and its outline suggests an old mine. At the back are several ultramafite layers. The hole fronts deep water, even at low tide, and is well protected from the elements. It would have made an ideal loading site. However, elsewhere on this peninsula the dark gneiss is in the intertidal zone, a setting that Frobisher's party seemed to avoid.

Better evidence of mineral exploitation was seen in the south peninsula. Here mining was confined to a length of 220 m, with two small trenches in the extreme north of the peninsula and a number of shallow scrapings and five spoil heaps southward (Fig. 27). The greatest elevation in this interval was 6.6 m above high tide. Ore, mainly 'black', was seen in loose pieces along the whole length of the peninsula, but contemporary accounts also noted 'red' (possibly garnetiferous or weathered ultramafite) and 'mixed' (possibly hornblendite-rich mafite). Banded 'black ore' almost identical to the Smerwick cobbles, is very common here (Fig. 28).

Trails of 'black ore' suggest the rock was loaded from both north and south peninsulas. Ore was carried in specially woven baskets,[8] but wheel barrows were also available. In August, 1578, the flats, to the east of the ridge of ultramafite must have been a veritable highway. It is interesting to note that the *Emanuel*, which had taken 30 tons from the mine, was debited for a 'druge',[8] a narrow hand-drawn cart, possibly used to transport ore.

Winter's Furnace Mine, Newland Island

A partially filled trench on Newland Island was discovered in 1990. It was 18 m long and 2 m wide and was thought to represent Frobisher's tiny Winter's Furnace Mine. The trench, in unconsolidated material, was aligned approximately parallel to the strike of nearby biotite gneiss and, therefore, like mafite at Countess of Warwick Mine and ultramafite at Countess of Sussex mine, 'black ore' (mafite) at Winter's Furnace appeared to be parallel (or 'conformable') to the grain of the rock. It was here that 5 tons of ore were mined in 1578, loaded into the *Armonell* and shipped to England. Remains of a small heap of 'black ore' were found just east of the trench. Two loose coal deposits 100 m northwest were rediscovered by Hall in 1861 (Hall 1864 **2**: 76-80) and may mark the location of an assay furnace operated in 1578. A plan of the mine was given by Hogarth (1990: Fig. 3) and a photograph shown by Fitzhugh (1993a: Fig. 6.7).

Possible mines of 'black ore'

Two occurrences, that must have been known to Frobisher's group are Tikkoon and Napoleon Bay (Fig. 29). In both cases 'black ore' and associated dark gneiss are surrounded by light-coloured rock and, for some distance seaward, are easily identified. Thus 'black ore' at Tikkoon can readily be seen from Kodlunarn Island, a kilometer to the west, and 'black ore' at Napoleon Bay is easily visible from Kamaiyuk Peninsula, four kilometers south-south west.

As would be expected from intertidal rocks subject to the ravages of wave and ice, neither occurrence shows evidence of Elizabethan prospecting or mining. Mines at these localities are not pinpointed in contemporary accounts.

The *Tikkoon* 'black ores' front the sea on the west side of a peninsula. At low tide, they can be traced as a series of auxiliary peninsulas and islets for over 500 metres. At high tide these occurrences are all but submerged. For example, the main occurrence (Fig. 29) at low tide is a peninsula 100 m long, 30 m wide and 23 m above the sea at its summit; at extreme high tide the rock is 3 m long and projects barely 1 m above the water.

Here, 'black ore' involves a number of rock types (Gonciar 1993), some containing considerable white calcite in layers parallel to a foliation. A well layered type, with alternating dark (hornblende) and light (diopside) fractions, is common. These layers contain isolated and attenuated blocks (xenoliths) of garnetiferous gneiss and diopside - bearing marble.

The 'black ore' is separated from biotite gneiss to the east by dark (hornblende-rich) gneiss, about 25 metres thick. The whole assemblage strikes north - northwest and dips westerly at 65 to 75°. It is cut by narrow dykes of a coarse-grained, pink feldspar-quartz-black mica rock (granite pegmatite).

The *Napoleon Bay* occurrence, appearing as a series of huge, black angular blocks and large dark boulders, is easily visible from the sea, at a distance of at least 4 km. These loose rocks have been dislodged from a sea scarp and, at low tide, are exposed for about 15 m along the coast. The rock contrasts markedly with the yellow sand and buff gneiss on its sides. Owing to the oblique truncation of the 'black ore' layers by the coastline, the width of the ore is only 6.2 m (Fig. 29) and, because of the dip of the rock, the true thickness decreases to 5.8 m. The 'black ore' lies between dark (diopside-hornblende) gneiss below (to the northeast) and light (garnetiferous) gneiss above (to the southwest).

'Black ore' on the west contains small lenses of maroon garnet in hornblende but this garnet gives way to fine-grained spinel (visible only under the microscope) on the east.

The 'black ore' is cut by narrow quartz veins containing minor white feldspar, yellow cummingtonite and pink apatite. On the west side of the occurrence, particularly near the quartz - feldspar veins, are patches of grey rock, composed principally of cummingtonite but containing some anthophyllite, biotite, feldspar and hornblende. Larger occurrences, with somewhat similar mineralogy, in Cumberland Peninsula, 270 km NNE of this locality, were described by Kamineni *et al.* (1979).

Mines which can be positioned approximately only

Four mines, *viz.* Best's Blessing, Countess of Sussex Island (Beare Sound), Fenton's Fortune, and Denham's Mount (Diar's Sound), have not been rediscovered in recent years.

Best's Blessing can be placed on the southeast side of Resolution Island ("amongst the Ilandes, which lye in Hattons headland in a sounde .. which seemed an indifferent place to harborough in"). Here Captain Best "founde a great black Iland" with an abundance of 'black ore' of similar appearance to that of Countess of Warwick Mine. Hattons Headland was said to be "the highest lande in all the straites [Hudson Strait]" and "about fifteen leagues" from Kodlunarn Island (G. Best in *SM* **1**: 111, 112, 115, 116), agreeing with the location on current topographic maps and the sketch in Best's book (*SM* **1**: 107). Prints in the National Air Photo Library do not reveal an obvious location of the "black Iland".

Sussex Island, locus of the largest mine in *Beare Sound*, was 9 leagues southeastward from Kodlunarn Island (E. Sellman in *SM* **2**: 64). Beare Sound appears to have been one and the same as Lupton Channel, through which Christopher Hall and Martin Frobisher rowed on August 15 and August 24 and the *Emanuel* of Bridgwater sailed on September 3, 1578 (G. Best in *SM* **1**: 121; T. Wiars in *SM* **2**: 253). The mine itself appears to have fronted a poor anchorage, unprotected from the north and northwest and situated in the southwest entrance to this channel (G. Best, *SM* **1**: 121-2; E. Fenton in Kenyon 1981: 201). A helicopter and ground search on August 2, 1992, disclosed solid coarse amphibolite and friable dark gneiss (but no typical 'black ore') at Lefferts Island, Matlack Island, an island southwest of Matlack, and the western shore of Loks Land.

Fenton's Fortune "at the entrance of Countess [of Warwick] Sound to the Eastwards", was small and situated somewhat inland. "Master Denham ... sayeth he cannot see how 40 tunnes will there be had, and that with great travayle to bring yt to the sea side" (E. Sellman in *SM* **2**: 65). Fenton described the ore as "black mixed with a white stone like the finte, the vaine wherof cometh from the Maine and lieth est and west" (Kenyon 1981: 190-191). This area [immediately east of Sharko Peninsula] was searched on the ground (1991) and by helicopter (1992) but neither mine nor 'black ore' was located.

Diar's Sound was obviously protected waters. According to Thomas Ellis (*SM* **2**: 43) it held "a calme and still water", and Fenton tells us (Kenyon 1981: 191) the "harbour for shippinge" was "verie good". It intersected "the souther land of the Countess [of Warwick] Sound" (E. Sellman is *SM* **2**: 65) and was on the east side of the sound (G. Best's map, *SM* **1**: 107). 'Prooffes' [assays] made in this area (E. Fenton in Kenyon 1981: 197) suggest that a furnace may have been set up on the adjacent land. They may be represented by a coal deposit on Ekkelezhun, a point of land (at Diars Passage of Frobisher) which protected the inner waters of Victoria Bay, "one of the safest and finest harbours" that Charles Francis Hall had ever seen (Hall 1864 **2**: 156). Kenyon (1981: 196), probably correctly, equates Victoria Bay with Diars Sound. This locality was revisited by the Smithsonian expeditions of 1990 and 1991: the unconsolidated soft coal deposit measured 22 m long, 9 m wide and up to 35 cm deep. It was intermixed with a very small amount of yellow European flint but Hall had also reported a sizeable wood chip, which he concluded was axe-hewen from a large oak timber (Hall 1864 **2**: 157-8).

The mine itself, *Denham's Mount* from which 260 tons were shipped in 1578, must be nearby. On August 18, 1578, Fenton "fownde iiii sortes [of ore] whereof to holde in goodnes equall in effect with any thother [of the other] mynes before discovered" (Kenyon 1981: 195). The mine may be covered by scree of a post-Frobisher landslide (talus is omnipresent here). However, at tidewater, intermittent exposures of a coarse-grained, hornblende-rich rock, 200 m north of the coal, may represent Frobisher's 'mixed ore'.

In our detailed search of this area, no trace of a stockpile or assay furnace was found. This was indeed puzzling and suggested that the coal at Ekkelezhun was *not* stored here as fuel for assay furnaces. It is unlikely that the deposit represents ballast, which would have required hauling coal 20 m above tidewater and depositing it purposely between two knolls. Possibly the coal was set aside to be used in a fourth voyage.

Doubtful mines

An opening in the sea scarp of Baffin Island, about a kilometer northwest of Kodlunarn, was described and illustrated by Kenyon (1975b: 144-5) as a possible mine site. The opening is 3 m high, 2 m broad and extends into the cliff for 1.5 m. It has all the indications of a sea cave: located within a metre of normal high tide and running parallel to a well developed fracture pattern in an easily eroded rock (coarse black, poorly foliated gneiss, composed mainly of hornblende, diopside, biotite and plagioclase). There is no evidence of former mining and no remains of excavated material in the immediate vicinity. The coarse-grained fraction of this rock (Hogarth *et al.* 1985, type *3d*; Hogarth *et al.* 1993 type *C5*) has now been removed from the listing of 'black ores' but has been dated as a country rock (this monograph, Appendix 2, specimens K/10B, K/10C).

The Lefferts Island occurrence (Fitzhugh 1993: 108-9) was examined in 1992. It is now considered to be a cove, the result of natural erosion of layers (hornblende gneiss and hornblende-almandine gneiss) more easily degraded than the adjacent rock (biotite gneiss).

Classification of 'black ores'

One hundred and twenty three samples of 'black ore' have been studied.[9] They have been classified according to mineralogical composition (Table 8). There are 5 rock groups and 19 types of dominant mineral association. With the exception of *D1* (from Dartford), all rock types representing the Dartford and Smerwick suites have known counterparts in Baffin. Hornblende is the link between all types: it is abundant in *A*, *B* and *C*, but comparatively rare in *D* and *F*. This mineral is coarse grained (commonly with grain diameters 0.5 to 2 mm), polygonal and shows no trace of derivation from earlier minerals. Chemical compositions are given in Table 9. The two most commom rock types are described below; their appearance under the microscope is shown in Figure 30.

Chemical composition of *B1* and *C1* ores (40-47% SiO_2, 13-23% MgO) are distinct from those of the surrounding biotite gneiss (70-75% SiO_2, ½-2% MgO). There is no obvious migration of elements from the 'black ores' into the wall rock,[10] nor increased metamorphism at the contact.

Type B1: hornblende, olivine, pyroxene ultramafite

This type (*B* of Hogarth & Roddick 1989) comprises 12 of our 17 'black ore' cobbles from Smerwick Harbour, but it is also common at Countess of Sussex Mine (12 specimens studied) and is found at Dartford (2 specimens studied). In hand specimen, most of the samples are layered due to bands alternately rich in diopside and hornblende (Fig. 28). Under the microscope, the hornblende has a distinctive brown (tan) colour. Crystals of olivine (forsterite) are dispersed in the hornblende layers. In addition, a special suite of minor minerals and significant quantities of chromium set this type apart from all others. Compared to ultramafites world wide, our rock is high in alumina (> 5%) and is low in magnesium (Mg/(Mg + Fe) = < 0.60, atomic).

Type C1: hornblende, enstatite mafite

This type characterizes our Dartford suite, accounting for 12 of our 22 specimens, but it is also common at Kodlunarn Island, where it occurs as 9 specimens (representing both Ship's Trench and Reservoir Trench). The rock is homogeneous, with a faint layering

Table 8: *Classification of 'black ores'*

Group, type and mineral association	l*	n#
A. Hornblendite + pyroxene association		
A1. Hornblende + diopside	I,E,B	21
A2. Hornblende + diopside + enstatite	E,B	5
A3. Hornblende + diopside + biotite	B	5
A4. Hornblende + enstatite	B	1
A5. Hornblende + diopside + enstatite + spinel	B	1
A6. Hornblende + enstatite + spinel	B	3
A7. Hornblende + enstatite + biotite	B	1
B. Hornblende + forsterite association		
B1. Hornblende + forsterite + diopside + enstatite	I,E,B	26
B2. Hornblende + forsterite + spinel	B	1
C. Hornblende + plagioclase association		
C1. Hornblende + enstatite + plagioclase	E,B	27
C2. Hornblende + biotite + plagioclase	B,I,E	8
C3. Hornblende + spinel + plagioclase	B,E	6
C4. Hornblende + diopside + plagioclase	E,B	5
C5. Hornblende + diopside + enstatite + plagioclase	E,B	2
C6. Hornblende + diopside + almandine + plagioclase	B	1
C7. Hornblende + enstatite + almandine + plagioclase	B	1
C8. Hornblende + enstatite + spinel + plagioclase	B	2
D. Diopside-rich rock		
D1. Diopside + forsterite + spinel	E	1
D2. Diopside + enstatite + spinel	B	4
F. Cummingtonite-rich rock		
F1. Cummingtonite + biotite + plagioclase	B	3

* locality: B Baffin, E Dartford (England), I Smerwick (Ireland)
# Number of specimens

Table 9: *Analyses of selected components (with standard deviation) for specific rock types*

rock	--------- Enstatite mafite ------		------- spinel mafite -------		--------- selvage ---------	
type	C1	C1	C3	C3	C2	C2
suite	KI	E	KI	E	KI	E
n	5	12	5	1	3	1
SiO$_2$(%)	45.6 (1.3)	43.73 (0.80)	37.9 (4.0)	45.38	45.7 (1.9)	45.17
TiO$_2$	0.37 (0.14)	0.27 (0.04)	0.41 (0.07)	0.34	0.55 (0.09)	0.18
Al$_2$O$_3$	12.0 (2.7)	14.19 (0.91)	17.4 (2.6)	17.11	14.0 (1.4)	15.86
Fe$_2$O$_3$	11.7 (0.36)	11.97 (0.50)	13.2 (1.8)	11.30	13.2 (0.8)	12.67
MgO	14.4 (1.2)	14.06 (1.24)	15.6 (1.3)	13.52	11.1 (1.1)	10.45
CaO	10.8 (0.8)	10.47 (0.47)	11.2 (0.4)	10.29	9.37 (0.17)	10.62
Na$_2$O	1.55 (0.29)	1.65 (0.16)	1.33 (0.60)	1.64	2.08 (0.46)	2.35
K$_2$O	0.55 (0.20)	0.33 (0.18)	0.26 (0.10)	0.30	1.91 (0.34)	1.13
Total	97.0	98.67	97.3	99.88	97.9	98.43
P (ppm)	100 (30)	< 50	< 50	50	< 50	50
Nb	5 (2)	< 5	5 (2)	< 5	6 (0)	< 5
Zr	39 (22)	10 (4)	20 (7)	18	41 (10)	16
Cr	289 (191)	553 (83)	209 (75)	167	199 (64)	206
Ni	413 (85)	444 (66)	540 (71)	542	319 (99)	337
Au (ppb)	2 (0)	n.a.	6 (3)	n.a.	5 (2)	n.a.
Ag	n.a.	n.a.	74 (30)	n.a.	48 (4)	n.a.
Pt	10 (5)	n.a.	6 (3)	n.a.	5 (0)	n.a.
Pd	2 (1)	n.a.	4 (0)	n.a.	2 (0)	n.a.

(**table 9**, cont.)

rock	----------- diopside ultramafite -----------			-------------- Forsterite ultramafite --------------------		
type	A1	A1	A1	B1	B1	B1
suite	KB	WK	TK	CS	I	E
n	4	1	6	12	12	2
SiO$_2$(%)7	40.7 (0.1)	40.83	41.8 (1.3)	45.4 (1.7)	44.0 (1.96)	43.50 (3.78)
TiO$_2$	3.36 (0.08)	3.37	3.75 (0.45)	1.27 (0.38)	1.70 (0.23)	1.46 (0.70)
Al$_2$O$_3$	7.22 (0.33)	6.95	7.33 (0.49)	5.86 (1.52)	7.72 (0.42)	7.10 (3.26)
Fe$_2$O$_3$	16.4 (0.1)	17.03	17.5 (1.0)	12.8 (0.8)	14.11 (0.71)	13.9 (2.72)
MgO	11.5 (0.3)	10.29	11.1 (1.5)	19.8 (3.1)	18.42 (0.89)	18.71 (0.30)
CaO	14.25 (0.26)	15.95	12.7 (25)	10.75 (1.9)	10.23 (0.83)	10.37 (1.99)
Na$_2$O	1.29 (0.08)	2.49	1.68 (0.18)	0.47 (0.23)	1.06 (0.27)	1.11 (0.45)
K$_2$O	0.93 (0.09)	0.08	1.07 (0.80)	0.28 (0.13)	0.16 (0.05)	0.17 (0.11)
Total	95.6	97.71	96.9	96.6	97.40	96.38
P (ppm)	1700 (90)	2100	1900 (480)	440 (220)	790 (390)	850 (30)
Nb	48 (2)	78	76 (10)	17 (6)	21 (6)	17 (10)
Zr	208 (10)	216	263 (39)	76 (26)	113 (16)	88 (42)
Cr	520 (54)	545	786 (86)	1470 (337)	1304 (161)	1422 (781)
Ni	382 (26)	254	320 (99)	850 (270)	764 (117)	683 (30)
Au (ppb)	10 (3)	14	8 (4)	5 (5)	n.a.	n.a.
Ag	n.a.	50	n.a.	n.a.	n.a.	n.a.
Pt	14.5 (3.5)	n.a.	< 10	< 5	n.a.	n.a.
Pd	2.5 (0.7)	n.a.	5 (0)	5 (2)	n.a.	n.a.

Abbreviations: CS Countess of Sussex Mine; E Dartford; I Smerwick Harbour; KB boulders, Kodlunarn Island; KI Countess of Warwick Mine (Kodlunarn Island); n number of analyses processed; n.a. not analyzed; TK Tikkoon; WK Kenyon Specimen (Kodlunarn Island).

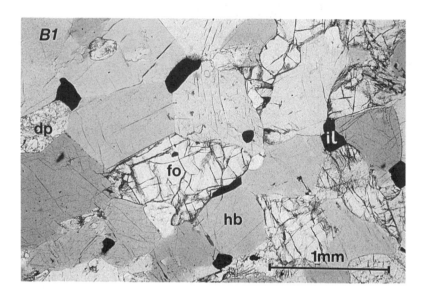

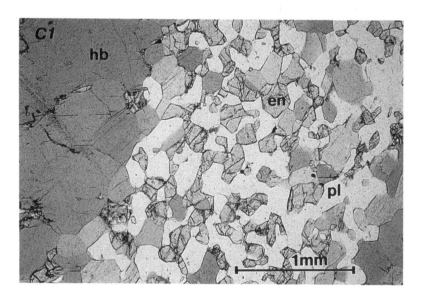

Figure 30: Typical 'black ores' viewed under the microscope. Bar scale is 1 mm long. Mineral constituents: dp diopside, en enstatite, fo forsterite, hb hornblende, il ilmenite, pl plagioclase. Sample localities: *B1* - North peninsula of Countess of Sussex Mine (Baffin Island), *C1* - Priory Road (Dartford, England).

detectable microscopically. The main mass is coarse-grained hornblende (average diameter 1.2 mm), which is green in thin section. Feldspar lenses (detectable under the microscope) and minerals within the feldspar lenses (enstatite, feldspar and the remainder of the hornblende) have an average grain size 1/4 that of the surrounding hornblende. The feldspar is plagioclase, extremely rich in calcium. Another notable feature of this rock is an almost complete lack of opaque minerals. Pyrite and magnetite, together, rarely exceed one or two tiny grains per microscope slide. Compared with mafites elsewhere, *C1* ores are high in alumina (> 10%) but low in silica (< 46%). At Kodlunarn type *C1* is associated with *C3* in the cores of boudins and layers, which are commonly rimmed with a selvage of *C2* (Fig. 24). Other specimens, tentatively placed with *C1*, are from Winter's Furnace, Kamaiyuk, Diana Bay and Willows Island; they are similar but differ somewhat in chemical composition, mineralogy and texture.

Chemographic analysis

Hogarth & Roddick (1989) compared specimens from Smerwick Harbour and Dartford with those of Baffin and Kodlunarn Islands. However, the equivalence of these suites was rather tenuous as only one sample from Baffin and one from Kodlunarn were used in the comparisons.

The conclusions are now strengthened with data from 11 additional specimens of type *B1* from Countess of Sussex Mine and 8 specimens of type *C1* from Kodlunarn Island. Type *C1*, probably the most common 'black ore' on Kodlunarn, is considered for the first time. The 5 Kodlunarn *C1* specimens of Table 9 were collected from bed rock.

In the present analysis, aluminium, chromium, iron, magnesium, nickel and titanium, will be used as discriminants. The range of composition of specimens of *B1* and *C1* from Baffin and Kodlunarn Islands will be established first. Then, comparison will be extended to specimens from Smerwick Harbour and Dartford.

Whole-rock analyses were examined with respect to aluminum, silicon, magnesium, chromium, nickel and iron (Fig. 31), which together make up 82 to 83 wt % of the metals in these rocks. For major elements, we have plotted the atomic ratio Mg/(Mg + Fe), the 'magnesium number', against atomic Al/(Al + Si). The *B1* and *C1* compositions of the Baffin rocks define two distinct fields which agree well with compositions of the same rock types in 'European' suites. The coherence of *C1* specimens is especially tight.

Chromium and nickel are the best trace-element discriminants. The *B1* and *C1* compositions define two widely separated fields and agree rather well with compositions of the equivalent 'European' suites. The Dartford *C1* ores are well clustered and give a surprisingly good fit with the Baffin *C1* domain.

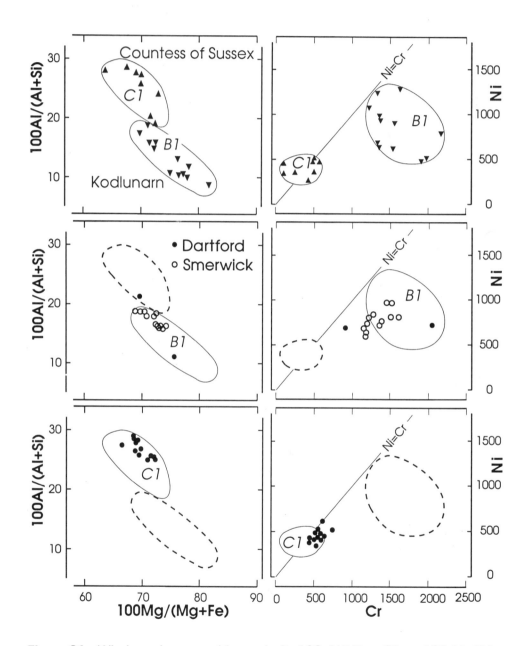

Figure 31: Whole-rock compositions. Left, 100 Al/(Al + Si) vs 100 Mg/(Mg + Fe), atomic; right, Ni vs Cr, ppm, wt. Upper, *B1* and *C1* ores, Baffin Island area; middle, *B1* ores, Smerwick Harbour and Dartford; lower, *C1* ores, Dartford.

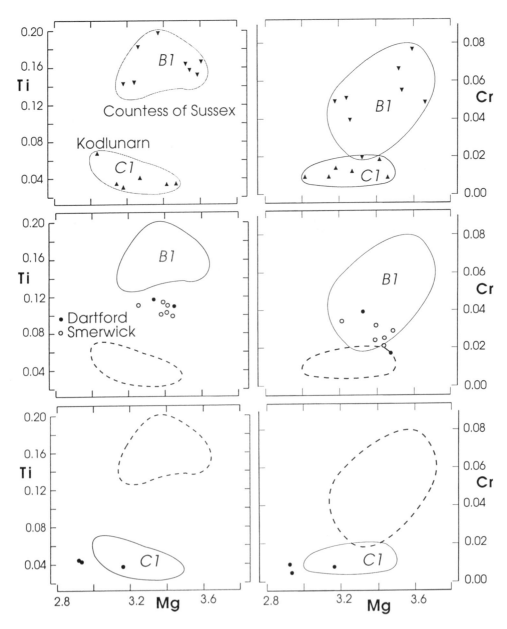

Figure 32: Hornblende compositions. Left, Ti vs Mg; right, Cr vs Mg; both as atoms per formula unit. Upper, *B1* and *C1* ores, Baffin Island area; middle, *B1* ores, Smerwick Harbour and Dartford; lower, *C1* ores, Dartford.

Figure 33: Glacially transported boulders ('erratics'), or blocks ice-rafted during post-glacial submergence, of type *A1* 'black ore' near the Reservoir Trench, Kodlunarn Island. They were possibly derived from Tikkoon Point, 800 m to the east.

For *mineral compositions*[11] we have used hornblende[12] and have plotted titanium and chromium against magnesium, as atoms per standard formula unit.[13] Here the coherence (Fig. 32) is not as good as in whole-rock compositions, especially for *B1* ores. However, there exists a general coherence of compositions within each of the two groups.

The coherence of *B1* and *C1* suites is now better established. We can say, that all the *C1* Dartford specimens probably came from Kodlunarn Island. The *B1* specimens, however, are not as tightly grouped; it appears they all came from a genetically related deposit in the Baffin Island area, but it is impossible to locate the exact source of the Dartford and Smerwick specimens.

Boulders of 'black ore' on Kodlunarn Island

Four boulders of 'black ore' (Fig. 33), one weighing at least 9 tonnes, are clustered about 10 metres south - southeast of the Reservoir Trench. They are very different from the indigenous 'black ores' at Kodlunarn Island. Minerals are segregated into black layers (hornblende, biotite, ilmenite) and green layers (diopside, actinolite), and the rock contains considerable white calcite in lenses which parallel the layering. Rock with similar external appearance was found in a specimen in the Kenyon collection from Kodlunarn Island and *A1* ultramafite from Tikkoon collected by Gonciar (1993; 4 specimens) and Hogarth (this research, 2 specimens).

The three groups (boulders - KB, Kenyon specimen - WK, and Tikkoon - TK) are compared, chemically, in Table 9. A glance at the table shows certain similarities within the three and, at the same time, differences from other rock types. They are relatively high in titanium and iron (due to the composition of hornblende and the presence of ilmenite) but, for the same reason they are lower in magnesium. The rather low totals are at least partly due to the presence of CO_2 (not analyzed but a constituent of calcite). Trace elements are even more significant. Elements in relatively high concentration are phosphorus (contained in apatite), zirconium (contained in zircon, a minor but omnipresent mineral) and niobium (contained in ilmenite).

We may conclude that the Kenyon specimen, labelled simply as "Kodlunarn Island, surface" was probably broken from one of the boulders. These boulders are more closely allied to ultramafite at Tikkoon than the indigenous rocks. They may have been deposited by a glacier during the Pleistocene or transported by ice during the post-glacial submergence.

Discussion

We may now reflect on the origin of our three suites of specimens: Baffin, Smerwick and Dartford. We have 24 specimens from Kodlunarn, of which at least 17 seem to be indigenous. However some, such as those sampled from the large boulders near the *Reservoir Trench*, are almost certainly transported. McGhee & Tuck (1993) suggest that these boulders may have attracted Frobisher's miners to the island. This is possible, but it is more likely that in July, 1577, 'black ore' was exposed on the sea scarp on the north side of Kodlunarn. Most pieces of ore in the old stockpile at the *Ship's Trench* are remnants of *in situ* occurrences nearby, but some may be "foreign to the island" (Roy 1937). This trench was probably finished in 1577.

At *Smerwick Harbour*, many 'black ore' cobbles belong to one rock type (*B1*), perhaps native to Countess of Sussex Mine, but three are variants of *A1*, with unusually high concentrations of vanadium (two specimens were described by Hogarth & Roddick 1993); they cannot be matched with any samples in our collection and perhaps represent the *Emanuel's* lading at Sussex Island (Beare Sound) or Denham's Mount, neither of which can be located at present. Two black sand samples contain an assortment of dark grains. Their ultimate sources can be placed as follows: Baffin Island area (magnesian ilmenite, diopside; possibly low-titanium magnetite), local Irish rocks (chamosite; possibly rutile, pseudobrookite), unknown (titaniferous magnetite, magnesio-chromite).

At *Dartford*, 11 of our *C1* specimens were collected from a well confined space.[14] The fact that these pieces were amassed together at a single locality adds credibility to a previous conclusion (Chapter 4) that ore from the second voyage was kept separate from that of the third and held in esteem during the period of operation of the metallurgical plant (1578-9). In the second voyage, Countess of Warwick Mine (Kodlunarn Island) supplied all the 'black ore'. In the third voyage, this mine supplied less than 6% of the total ore landed in England (see Table 4). It was shipped in three barks, in each case in quantity smaller than other ores and, after unloading at the docks, cart transport into Dartford, and storage at the Manor House, all ores would have been thoroughly mixed. We may conclude that at least half our Dartford suite represents lading from the Ship's Trench in the second voyage.

Geochronology

The equivalence in age of two specimens of 'black ore' from Smerwick Harbour with one from Kodlunarn Island has been reported by Hogarth and Roddick (1989). As was pointed out by these authors, isotopic ages place a stronger constraint on the source of samples than do petrology and chemistry alone. By way of endorsement of this view, potassium-argon ages have been determined on a further 18 of the petrologically classified specimens, 8 from the Baffin Island area, 6 from Dartford and 4 from Smerwick Harbour (Fig. 34; Appendix 2). A dolerite (diabase in North American terminology) cobble from Smerwick Harbour was also dated.

It is evident that most whole-rock ages of the additional specimens of 'black ore' are consistent with previously published hornblende ages, viz 1700 - 1900 million years (Ma). These ages are, as pointed out by Hogarth and Roddick (1989), typical of the ages of high-grade metamorphic rocks from SE Baffin Island.

We are able on the basis of new data, to strengthen the case for geochronological coherence of the three groups of specimens. It is evident from Figure 34 that a second group of ages is present in 'black ores' from Dartford and Smerwick Harbour, *viz.* whole-rock potassium-argon ages of 1350 - 1600 Ma. These same ages appear in specimens of skarn and gneiss from Kodlunarn, and Baffin Islands and are explained by the presence of microcline and/or biotite, which may release argon even at moderate temperatures.[15] Similar ages have been previously recorded from rocks nearby, for example by Wanless *et al.* (1968: 62-4) who list four 1500-1555 Ma-old granites and gneisses in southeast Baffin Island. It should, however, be noted that samples 88/4 (type *B1*) and E23/2 (type *D1*), which contain neither mica nor feldspar, also yield these 'young' ages and may, in truth, express a later igneous or metamorphic 'event'.

Our samples, therefore, belong to two distinct groups, which give mean ages (standard deviations in parentheses) of 1470(90) Ma for 10 samples and 1840 Ma(90) for 11 samples. These two age groups may leave unanswered questions concerning interpretation of the geology of Kodlunarn and adjacent Baffin Island. They do, however, fortuitously afford a further demonstration of the affinity between the three sample suites, in that similar whole-rock age populations characterize all three.

A specimen of dolerite (88/11A) was collected from Smerwick Harbour, as a mafic rock possibly related to our mafic - ultramafic suite. However the age (36.9 Ma), and marked rhythmic zonation of plagioclase and diopside, mitigate against this hypothesis. The mineralogy and age are similar to dolerite from the core of an exploration hole drilled just west of Ireland (Seemann 1984) and a dyke at Ballinrannig, 2.5 km southeast of Dún-an-Óir (Horne & MacIntyre 1975).

Origin of the 'black ores'

Mineralogy and bulk-rock chemistry of 'black ores' suggest metamorphic derivation from the deep crust or upper mantle. For example, the occurrence of olivine (forsterite), chrome spinel,[16] enstatite and diopside, and a small but significant nickel-chromium abundance in the bulk-rock composition, point to a deep source. There is nothing in the

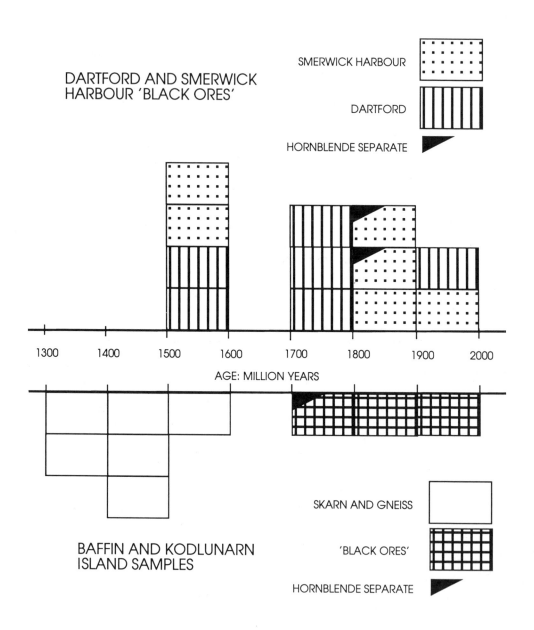

Figure 34: Potassium - argon ages of 21 samples of 'black ores' and country rocks.

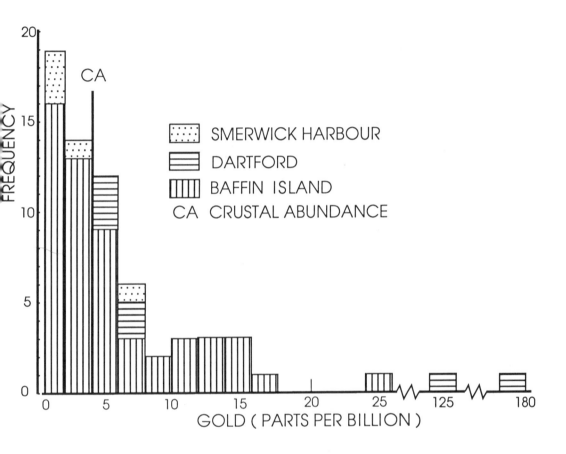

Figure 35: Histogram showing gold content of 66 samples of 'black ore'.

present data to suggest these rocks were not initially magmatic. However, only five of our specimens (group D) have mineral compositions characteristic of magmatic rocks. Hornblende, the principal constituent of the other 'black ores', is a typical metamorphic mineral. Therefore, if we assume an igneous origin, the rocks have undergone considerable changes since crystallization. Repetitive layering, characteristic of group A and B 'banded ores', may be a compositional remnant of an original igneous layering. The lack of migration of elements from the 'black ore' into the wall rock, the fractured but otherwise unchanged biotite gneiss at the contact, and the very coarse-grained and pristine nature of the hornblende, suggest the 'black ores' were injected as cooled solids into the wall rock and that metamorphic transformations had taken place previously at depth.

As material rich in magnesium, iron and silicon migrated upwards, it slowly cooled and assimilated constituents (including water) from the wall rocks. This would explain the generation of hornblende, a hydrous mineral. Migration was along the avenue of easiest access, the natural grain or 'foliation' of the rock. As upward migration and cooling continued, chemical reactions and ion exchanges became more sluggish, and at about 700°C and 20 km depth[17] atomic interchange between neighboring minerals virtually ceased. Finally upward migration came to a halt altogether and the rocks were trapped several tens of kilometers below what was then the surface of the earth[18]. The time was early Proterozoic, 1900 million years ago. It remained for uplift and deep erosion to expose these narrow layers of rock or 'sills' to plain view.

Gold, silver and platinum

Gold was the centre of attraction in the Frobisher enterprise of 1577 and 1578. High gold contents were reported in some early assays (e.g. Table 5; bulk assays 1, 2, 5, 6). Whereas, in most early tests, silver exceeded gold in monetary value (Hogarth & Loop 1986), gold always commanded an attention that silver lacked and, for the most part, silver was ignored by the Frobisher group.

Now, after a lapse of 400 years, it was felt time to check, with modern methods, precious metals in Frobisher's ores. Like the Elizabethan assayers, our attention centred on gold, but for a different reason: in low-grade ores (and preliminary tests showed they were indeed low grade), accurate determinations of gold are much more easily obtained than determinations of silver. [19] In the present study, 66 samples of 'black ore' were analyzed: 54 from Baffin Island and vicinity, 7 from Dartford, and 5 from Smerwick Harbour. With the exception of 2 specimens from Dartford, gold was *extremely* low (Fig. 35). Half our specimens contained less gold than the average of the Earth's crust (*viz.* 3.5 ppb; Li & Yio 1966). Of the common 'black ore' types, *A1* appeared to be somewhat higher grade than others and approached the mean value of ultrabasic rocks or ultramafite (*viz.* 11.4 ppb; Boyle 1979).

Similar, low values were obtained from the country rocks. For example, five samples of mafic gneiss associated with 'black ore' at Countess of Sussex Mine averaged 1½ ppb, three from the sea cave on Baffin near Kodlunarn averaged 3 ppb gold; two samples of biotite gneiss at the foot-wall contact of 'black ore' on Kodlunarn gave 2 ppb, two from central Kodlunarn gave 2½ ppb gold.

Two specimens from Dartford did show significant gold: E 23/2 (180 ppb Au) and E44/19 (125 ppb Au)[20]. The former held the only *D1* assemblage in our suite and the only specimen in which the arsenides gersdorffite and niccolite were observed. Presumably the gold was present as native metal inclusions in the arsenides. Specimen E44/19, an amphibolite (type *C1*), contained a small amount of pyrite (with 5.4% Ni) through which gold was distributed inhomogeneously.

The low content of precious metals in our suite may be linked to the dearth of sulphides and arsenides into which these elements would partition at high temperature. 'Red ore' from Jonas Mount was not available for our research. It had been in short supply, even during the voyages. It may have represented an iron-rich capping over weathered sulphides, a gossan, in which case the high assay obtained by Burchard Kranich (40 oz Au/T or 1200 ppm) was possibly correct. Gold and silver are known to concentrate during weathering, in some instances resulting in rich soils and gossans, even in northern Canada (Boyle 1979: 431-45). 'Red ore' from Jonas Mount, a loose sand, is different from 'red ore' from Countess of Sussex Mine, solid rock.

Let us now consider the early gold proofs of 'black ore' (Table 5) and pose the question: where did the Elizabethan assayers go wrong? The highest values certainly belong to the earliest assays (Hogarth & Loop 1986), but it is not clear whether the gold was completely 'parted' (or separated) from silver, nor are we given details of analytical procedure. After assay No. 5, tests in England were made before a delegation of commissioners and the gold content of the ore dropped immediately from 13 ½ to 2 ½ ounces per ton. One may suspect mischief in the early assays, but the later, supervised tests are also difficult to explain. The latter report a gold content of about 1 ½ ounces per ton, 2000 to 46,000 times the content of 64 of our samples, and 200 to 300 times the non-conformist 'black ores' E23/2 and E44/19. In the Elizabethan assays at Dartford, galena (elsewhere associated with appreciable gold) was a major component of the crucible charge, but it is hard to believe that Shutz, reputedly acquainted with metal-lurgical technology of the day, would not have tested his additive for precious metal and made allowance for it. Likewise silver, known to be present in small but significant amount in the Caldbeck additive (see Chapter 4), should have been determined and deducted from the assay.

Platinum, known to concentrate in some mafic and ultramafic rocks (e.g. Barnes & Duke 1990; Boyle 1982) would, like gold and silver, remain in the precious metal bead after cupellation (near the final stage of fire assaying). This metal was first described scientifically in 1750 and, therefore, the Elizabethans may have confused it with gold or silver. Accordingly, new analyses were made of specimens from the "payzone" and stockpile of Countess of Warwick Mine but, alas, platinum was present in the order of 5-10 ppb, about 1/5 to 1/10 its crustal abundance. Analyses of specimens from the ultramafic "payzone" of Countess of Sussex Mine also produced low values (Table 9).

The inescapable conclusion is that either the analysts were incompetent or that Jonas Shutz, Robert Denham, and perhaps others, added gold-bearing material to the charge deliberately. However, what sustaining advantage the assayers hoped to gain in fraud, is difficult to imagine.

Notes

1. Roy (1937: 34-7) described 'amphibolite' as composed of hornblende (80% vol.), pyroxene (10%), and feldspar, titanite, calcite, mica and iron-titanium oxides (together 10%). 'Pyroxenite' was composed of augite (50%), hornblende (15%), mica, feldspar and apatite (together 35%).

2. This same conclusion is repeated by Dr C.E. Tilley of Cambridge: "it appears that this bronzy lustred mica was the cause of the gold rush" (quoted by Roy 1937: 34). However, most of our specimens and many blocks of 'black ore' in the Priory wall at Dartford do not have obvious bronzy mica.

3. According to Blackadar (1967a) most bedrock on the northeast shore of Frobisher Bay belongs to the following units: quartz-feldspar gneiss, grey granite gneiss, and migmatite.

4. A search of the Cambridge Harker collection by Dr G. Chinner revealed two thin sections only (without the relevant hand specimens). No mica was observed in either section. They appear to be somewhat different from the Cambridge specimens described by Drs Illing, Priestley and Tilley.

5. The date of discovery is that of G. Best (*SM* 1: 64); D. Settle (*SM* 2: 17) gives Aug. 4.

6. Another possibility is that the washed fines held "golde plainly to be seene" (G. Best in *SM* 1: 64). The 'golde' may have been mica, naturally oxidized at the surface (cf Roy 1937).

7. In this monograph, a mafic rock (or mafite) is one that is composed largely of ferromagnesian (or mafic) minerals, such as hornblende, diopside, enstatite or biotite, but containing 3-10% plagioclase by volume. An ultramafic rock (or ultramafite) is one that is composed almost exclusively of mafic minerals, and containing less than 3% plagioclase. In the present usage the terms are applied equally to igneous and metamorphic rocks with apparent igneous parentage.

8. Lok's account, 1578. *HL*, HM 715, f. 9V. "paid to Geffrey Turvile and Rycharde Bowland and other officers in the Tower for ... Basketts viijᶜ at iijᵈ the peceli 10"

Sellman's account, 1579. *PRO,* E 164/36, f. 82. "The Ship Manewall is debter ... for j Druge [value not stated]".

In the final parting from the mine (August 31 1578) Edward Sellman remarked (in Kenyon 1981: 200) "I was forced to send my Pynnasse to the Countesse of Sussex Myne ... to burie divers things lefte there", but just what and where has not been ascertained.

9. These specimens were distributed as follows:

Kodlunarn Island	24
Countess of Sussex Mine	24
Dartford	22
Smerwick Harbour	17
Tikkoon	12
Napoleon Bay	8
Kamaiyuk	5
Diana Bay	4
Willows Island	4
Winter's Furnace	2
East Sumner Island	1
	123

Slabs of the Smerwick Harbour and Dartford specimens are stored in the petrology collection, Department of Mineralogy, Natural History Museum, registered as BM 1990, P2 and BM 1990 P3, respectively. Specimens from the Baffin Island area are stored in the petrology collection, Department of Geology, University of Ottawa.

10. Comparison of chemical compositions of ore and country rock at Kodlunarn (below) did not demonstrate migration of transitional elements from the 'pay zone' into biotite gneiss.

Rock	location	n	Fe%	Cr(ppm)	Ni(ppm)	Zn(ppm)
C1	'pay zone'	5	8.2	289	413	91
C2	foot-wall contact	3	9.2	199	319	156
gneiss	foot-wall contact	2	2.5	20	19	35
gneiss	central Kodlunarn	2	1.7	20	15	44

Chemical analyses (major, minor and trace elements, excepting Au, Ag, Pt, Pd) were made by X-ray Fluorescence Spectroscopy (XRF), some at X ray Assay Laboratories (Toronto), some at University of Ottawa.

11. Mineral analyses were made by wavelength dispersion spectroscopy with the electron microprobe (WDS-EM) mainly at the Mineral Sciences Division, Canadian Museum of Nature, but some were made at the Department of Mineralogy, Natural History Museum and at the Department of Earth Sciences, Carleton University.

12. The consistency of chemical composition of hornblende within an individual polished thin section is demonstrated by standard deviation. Examples of two specimens are: 91/73E (type *B1*, collected *in situ*, south peninsula, Countess of Sussex Mine, Baffin Island) and K5/12A (type *C1*, collected *in situ*, Ship's Trench, Countess of Warwick Mine, Kodlunarn Island). n = number of analyses considered. Standard deviations are in parentheses. calc = calculated value from stoichiometry.

Specimen	K5/12A		91/73E	
n	6		7	
SiO_2	45.37	(0.30)	44.21	(0.32)
TiO_2	0.32	(0.02)	1.54	(0.04)
Al_2O_3	11.83	(0.19)	12.53	(0.15)
Cr_2O_3	0.17	(0.03)	0.43	(0.07)
Fe_2O_3 calc	10.05	(0.11)	3.16	(0.05)
FeO calc	0.21	(0.01)	2.40	(0.04)
MnO	0.15	(0.01)	0.06	(0.02)
MgO	16.16	(0.15)	17.07	(0.12)
CaO	11.20	(0.11)	12.60	(0.07)
Na_2O	1.90	(0.01)	1.62	(0.04)
K_2O	0.35	(0.02)	0.51	(0.03)
H_2O calc	2.11		2.08	
Total	99.82		98.21	

F, Cl and Ni were below the detection limit.

13. The standard formula unit of hornblende is
$(Ca,Na,K)_{2-3}$ $(Mg,Fe,Ti,Mn,Cr,Al)_5$ $(Si,Al)_8$ O_{22} $(OH)_2$. Our calculations have been based on $Mg+Fe+Ti+Mn+Cr+Al+Si = 13$ atoms with 46 positive charges on these atoms and K, Na and Ca.

14. A precise location is the Dartford District Archaeological Group's excavation D44 (Phase 2), 1982, at the west end of the former pattern shop of J. Hall's engineering works. Frobisher ores were found in Trench 1, Layer 6. Approximate location: this monograph, Figure 17, locality 2.

15. The common silicate minerals can be ranked according to their argon retentivity and hence their suitability, or otherwise, for potassium - argon age studies. Dalrymple & Lanphere (1969) and Berger & York (1981) suggest the following ranking in order of decreasing argon retentivity: hornblende - muscovite - biotite - feldspar. Alkali feldspar and plagioclase (from metamorphic rocks) are regarded as unsuitable for K - Ar dating by virtue of their inability to retain argon even at room temperatures. On the other hand, hornblende is the most resistant of the common silicate minerals to argon loss.

16. At Countess of Sussex Mine, chromium varied from 3.17 (hercynite) to 50.73% Cr_2O_3 (chromite) in spinel-group minerals. Chromium was also detected in ilmenite and silicates. Generally, associated phases can be ranked on a descending order of chromium content as follows: spinel > hornblende > biotite > ilmenite > diopside $\geq$ enstatite > forsterite.

17. Some temperature determinations were: 720 (90) °C for ten 'black ore' specimens from Countess of Sussex Mine using the two-pyroxene geothermometer of Kretz (1982). Hogarth & Roddick (1989) gave temperatures of 620 to 740°C. A pressure of 6.1 (0.5) kB for four *B1* ores (two from Smerwick Harbour, one from Dartford, one from Countess of Sussex Mine) was determined, using the cordierite - spinel geobarometer of Seifert & Schumacker (1986) at an assumed temperature of 700°C.

18. Equilibrium temperatures of three country-rock mafic gneisses were essentially the same as the 'black ores' from Countess of Sussex Mine (viz 720 (30)°C), using the two-pyroxene geothermometer of Kretz (1982). Jackson & Morgan (1978, and Geological Survey of Canada Map 1475A) show this area to be a high temperature - high pressure ("granulite") terrane.

19. Gold was determined by several methods. The ones quoted here were made by neutron activation after NiS extraction or by direct current plasma spectroscopy after fire-assay extraction. The latter method was also used for the platinum and palladium determinations quoted in Table 9. Silver was determined by atomic absorption spectroscopy using a graphite furnace.

20. Analyses by neutron activation (no pre-concentration). Other specimens from Dartford (3) and Smerwick Harbour (5), not used in the compilation, but determined by this method, were below the detection limit for gold (40 ppb).

CHAPTER 6

Aftermath and Conclusions

The Frobisher expeditions were sent out, first, to find the Northwest Passage and test its feasibility as a route to the Orient, later, to discover and mine gold-silver ore and finally, as a subsidiary project, to establish an Elizabethan colony in the Arctic. Realization of these objectives fell well short of the mark. Search for the Passage and serious geographic exploration were discontinued after the first expedition. The voyages, which might have made a notable contribution to northern exploration, in truth, retarded the progress of geographical knowledge for at least 200 years.[1] The mines proved worthless and hope for a northern colony was abandoned. No money was made in this enterprise, in fact much was lost, and for some men, like Lok and Walsingham, the voyages led to bankruptcy. It is therefore not surprising that most researchers have concluded that the three voyages were a complete failure in all aspects. But before we reach a conclusion, the record should be examined impartially.

Exploration

Exploration was seriously hampered by the veil of secrecy imposed by the Privy Council, which enwrapped and concealed all geographical data. The contemporary accounts of George Best (*SM* 1: 1-129), Dionyse Settle (*SM* 2: 1-25) and Thomas Ellis (SM 2: 27-51) had all precise geographical data expunged. The maps in Best's book, ascribed to James Beare, were vague and misleading. The general map (*SM* 1: 2) places Frobisher's landfalls immediately north of 'Bacaleos', a name commonly identified with Newfoundland, and in the text there are several references to Meta Incognita being south of Labrador. It was only natural that Frobisher Straits should be placed on the east coast of Newfoundland. Thus, Anspach (1819: 369-70) identified the Strait with Catalina Harbour and the 'ore' with local pyrite ('catalina stone'). The error was repeated by Martin (1983: 6).

These charts, however, deserve more credit than has previously been given. The general map is an improvement on contemporary cartography. Labrador, at this date, was commonly confused with Greenland (Biggar 1932). Bacaleos was not located in Newfoundland but on the Ungava coast, and Frobisher's Strait (Frobisher Bay) and Meta Incognita (Baffin Island) were shown, correctly, immediately to the north.

The more detailed map (*SM* 1: 107) was also misleading. The Elizabethan navigators, trusting the Zeno map (which later proved to be fictitious), believed they had passed 'Friseland' and were, therefore, south of Greenland. This led cartographers, starting with Jodocus Hondius (Fig. 36), to transect Greenland with the Strait! Reflection on the

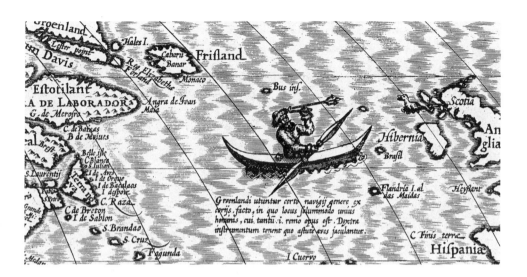

Figure 36: Part of a map of the Americas published by Jodocus Hondius the elder in 1606, showing 'Groenland' transected by Frobisher's Strait, Buss Island (*Bus ins.*), and Inuk-in-Kayak. The harpooning Inuk has been engraved in copper and faithfully copied from the first German edition of Settle and, therefore, printed in reverse. The Latin inscription under his boat, reads "The Greenlanders use a particular sort of boat made out of hides in which there is room for only one man who needs great strength to use the paddle. In their right hands they hold a device with which they cleverly shoot at birds". *NAC*, MNC 27653. Published with permission.

construction of Beare's maps would have shown that Frobisher did not visit Greenland in the position marked: Greenland was merely spotted as tradition dictated.

Although the error was suspected by Alexander Dalrymple (1789: iv), who equated Frobisher's Strait with Lumley's Inlet of John Davis, it remained for commander Alexander Becher (1843) to show, conclusively, the true location of Meta Incognita. Becher's prime source of information was Christopher Hall's ship's log of the first voyage. It is surprising that this exercise was not performed earlier, since the log had been available from the time of Hakluyt (1589: 615-622).

Several manuscript maps are now known that, regrettably, have remained hidden until the present century. The most comprehensive, but drawn in small scale and showing few place names, is the 'William Borough' chart at Hatfield House.[2] Also to be noted are John Dee's small manuscript map presented to Queen Elizabeth,[3] Robert Dudley's compilation in Munich,[4] and the silver medallion of the world, struck in 1580 (Christy 1900). A rare map, showing Frobisher's landfalls in their true position, was published by Hakluyt (1587).[5]

Although the immediate contribution of the three voyages was slight, they did stimulate future exploration and acquisition of land. Soon after return of the second voyage, Elizabeth, mindful of her tenuous sovereignty, had John Dee draft a formal claim to territory "discovered, inhabited, and partly conquered by the subjects of this Brytish Monarchie". In it was included Meta Incognita.[6]

Ethnology

Frobisher's voyages introduced Europeans to the Baffin Inuit and their culture. The first voyage returned to England with a male Inuk, the second with a man, woman and her infant son, but none lived longer than a few weeks.[7] Although the existence of the Inuit, their huts and kayaks were already known to Europeans,[8] this was the first detailed account of their culture and, for many, the first face-to-face encounter with the people. It created immediate fascination.

As a permanent record of the Inuit brought to England by the Frobisher expeditions, a Flemish portraitist, Cornelis Ketel, was commissioned to paint them in oils. He produced four pictures of the first Inuk, and the Dutch engraver-sculptor William Cure cast him in wax, presumably all accomplished post mortem.[9] Similarly, four paintings were made by Ketel of the woman taken in the second voyage.[10] Portraits by two other Dutch artists and a panorama of the skirmish at Bloody Point are described and reproduced by Hulton (1961, 1984), Hulton and Quinn (1964) and Cheshire *et al.* (1980).[11]

The appearance, diet, dwellings and habits of the Inuit soon became known to the European reader, particularly through the various editions of Dionyse Settle. The French, German and Italian editions contained somewhat fanciful illustrations, apparently inspired by the available paintings and text descriptions.[12] Jodocus Hondius, the elder, in turn, appears to have an Inuk-in- kayak from the German edition, for his map of America (Fig. 36).

The accounts and illustrations were, therefore, a milestone in European knowledge of Inuit culture in arctic America. Although Frobisher's interchange with the natives was stormy, lessons were learned which led to more peaceful encounters in the future.

Robert Denham

For most metallurgists and assayers, the sour reputation of the Frobisher gold venture hastened the end of their careers. Burchard Kranich fell into disgrace but did not outlive the Dartford enterprise. Jonas Shutz was regarded with suspicion and soon faded from the picture. Giovanni Battista Agnello was seen as a fraud and, in 1578, his association with the company was virtually terminated. Not so Robert Denham: he reappears, briefly, as a shining star five years after the closure of the furnaces.

Denham was one of the few major players who emerged from the Frobisher scene unscathed; in fact he profited from his association with the company. We see him consecutively as a goldsmith (London, 1560s)[13] as an assistant assayer (second voyage), chief assayer and mining consultant (third voyage), and as master assayer (Dartford, 1578-9). He was regarded as a competent technologist and impartial administrator by the Frobisher and Fenton cliques, and his services were sought by both parties. No doubt he acquired much useful experience during his association with the Cathay Company.

In 1579 Denham disappeared from the record but he re-emerged in 1584. Thomas Smythe, chief farmer of customs in London and a charter shareholder in the Company of Mines Royal, which was awarded a franchise to mine and produce metals in England and Wales, leased the royal privilege of this company on behalf of himself and others. From late 1583, Smythe and partners explored mineral properties in Cornwall, Devon and south Wales. Denham, representing William Burde, was sent to Piran Sands to appraise the Treworthie lead-silver mine, previously worked by Burchard Kranich and William Carnsewe (1555- c.1565) under the hardship of constant flooding.[14] With the advice of Denham, the mine was now dewatered and the vein followed laterally below the old workings for some 100 metres. The highgrade stockpile of silver ore, amassed in Fowey by Thomas Treffrey in 1559 for Burchard, was shipped to London and sold, a lead-silver mine was profitably reworked at St. Columb Minor, and copper mines were opened with some success at St. Anne, Illogan, St. Just and St. Ives. The copper ore was hand sorted at pit head and shipped to Neath in south Wales for blending with ore from Cardigan. Denham appears to have acted as consultant in all these activities. We last hear of him in March 1587, when the Cornish mines were being closed: he disappears, finally, from history.[15]

The impression gained is that Denham was a highly respected mining technologist and cannot be blamed for the demise of the mines. Judging from State Papers and the Carnsewe manuscript, the tenor of silver in these ores (on which success depended) varied widely and economy was marginal, at best.

Officers and seamen

On return from the voyages, officers and sailors became famous overnight, a fame that probably lasted throughout their lifetimes. They had delved into the unknown, forced their way through "mountainous yce", braved howling gales, penetrated thick fog and steered through treacherous tidal eddies. They had bartered with an unknown people of strange language and customs. They had painfully scaled a precipice to plant the Queen's banner in a far-off land. These accomplishments completely outshone the tarnish of economic disaster of the mining enterprise and failure to find the Passage. This was the age of awakening of overseas exploration. No doubt, the participants in the Frobisher voyages entered their experiences in their curriculum vitae, which ensured employment during their active careers. We have found 28 mariners of Frobisher's company mentioned in publications and manuscripts concerned with later voyages or naval service. Their exploits are summarized in Appendix III.

The post-Baffin accomplishments cannot, individually, be regarded as great. Rather they are links in a chain of events that led to the broadening of England's power and, to a lesser extent, expansion of geographical knowledge. Some mariners careers overlapped the military. Edward Fenton led a voyage of exploration to Africa and South America (1582-3), but he served with the army in Ireland (1566, 1579-80). He also commanded the *Mary Rose* against the Armada (1588). Likewise, Gilbert Yorke served in Ireland (with the navy 1579-80), against the Armada (1588), but sailed with Drake and Hawkins to Panama (1594). John Smythe of Bridgwater, the erstwhile master of the ill-fated *Emanuel*, commanded a bark of 70 tons against the Armada (1588). Luke Ward, Fenton's man in Baffin (1578) and Africa-South America (1582-3), commanded the small galleon *Tramontana* and 70 men against the Armada (1588).

Another interesting mariner, a native of Devon and captain of the *Hopewell* in Frobisher's third voyage, was Henry Carew. In 1575, he had been given letters of marque by Don Luis de Requesens, the commendator of Castile, to cruise against the enemies of King Philip and take any rebel ships (including English merchantmen), with cargo and mariners, found in Spanish waters. In 1579, he served under Valentine Pardieu, governor of Gravelines (near Calais), an administrator of doubtful loyalty. In January 1580, Carew "passed to Nantes and so to go by sea to Spain". In April of the same year, he is listed as a Catholic from Devonshire living in Paris and in June, Sir Henry Cobham (English ambassador, resident in Paris) notes that Carew [Carye] was made captain of a ship destined to Ireland. This report was confirmed by the mariner John Ley, who noted that "M[aste]r Carye tolde him he had bynn a Captaine w[i]th M[aste]r furbusher in his v[o]iage to Cathay, & being the King of Spaynes Godsonn, and not liking the Religion in England, he came into Spaine, and had got[t]en thentertainement of a captaine to serve the King".[16] Carew was told the action was planned against Portugal, but Ley was informed by one Oliver that this force was to follow-up on James Fitzmaurice's raid in Ireland. Carew was therefore appointed to transport troops and materials to Smerwick Harbour. One wonders whether, after this duty, he remained behind to defend Fort Dún-an-Óir and was later punished by the British. Certainly, he is not heard from again in the English State Papers.

James Aldaye is a rather special case. He had been a merchant and mariner since at least 1551. In 1575 he prayed that, should Michael Lok outfit a voyage for discovery of the northwest route to Cathay, he would be considered for service, and "therein to adventure life to the uttermost point" (Hakluyt 1598-1600 1: 412-3; 2: 319-20). Aldaye was duly included in the ship's company in the first voyage and paid £5 in advance, possibly at the top mariner's wage (25 s per month). In 1579, Aldaye's experience in the first voyage gained him a commission from King Frederick II of Denmark with an express purpose to strengthen the ties of Denmark with Greenland (Pingel 1845: 637). That summer he took two vessels and coasted southern Greenland, but ice prevented his landing. He had failed to re-assert Danish sovereignty over Greenland and to establish the state church (Lutheranism) on the island. In spite of this, the commission was renewed. In 1580, Aldaye was to lead a second expedition to Greenland, but when his ships were commandeered for defence in the Baltic, Aldaye was promised employment at the Danish court. He then disappeared from history (Gad 1971-82 1: 194-95; Pingel 1845: 637-64).

Whereas many officers from the northwest voyages went on to further exciting adventures on the seas, the same cannot be said for captains Randall, Courtenay and Newton. It may, therefore, not be amiss to describe the effects of the voyages on the later careers of these three men.

Randall had a multifaceted career. By profession he was a merchant (from 1553) and ship owner (from 1565 or earlier), with trade extending beyond the British Isles. He was frequently accused of complicity in piracy (from 1560). At the same time, he was well respected by his community and was, for several terms, appointed bailiff (1565-87).

After the voyages he was a hero in his private world, Weymouth, and was thus able to settle into legitimate trade. He successively held the posts of vice-admiral of Dorset (1583), mayor (1584-5), alderman (1586) and justice of the peace (1587). However, his role in piracy was not completely forgotten and we find him jailed for refusing to testify in a piracy case, but he was soon released (1580). He took no part in the Armada battles but was charged with fortifying Weymouth and vicinity against possible invasion from the sea (1587).[17]

Thomas Courtenay was of a different ilk. As a member of one of Devon's noble families he was brought up in privileged circumstances, though as ninth in line to Sir William Courtenay's estate, he stood little chance for inheritance. However, he married (1564) Alice Parker who fell heir to property in North Molton, Devon, whence the couple took up residence. Voyages, mercantile service, the military and, perhaps, piracy were digressions from the routine administration of family affairs. After compensation for his ship and services in Ireland (1579), he was released from duty with the ominous rider that should he not be paid, he might return to "hys old occupacon".[18] Then (1580) came the pardon, which perhaps referred to former piracies as well as the Dún-an-Óir incident (Chapt. 3). He was now cleared of all charges involving "murders, felonies, piracies, seizures, robberies and all other offences and evil doings whatsoever, committed upon the seas". A guaranteed protection against ensuing legal action completed the exoneration.[19] Courtenay re-appeared in Kingsbridge, Devon (1585) where he was involved, with sword drawn, in a local feud. The Courtenays and friends were charged with "aussaultt battery and affraye".[20] It seems that the 'voyages' had little effect on Thomas' activities.

Richard Newton of Bridgwater belonged to yet another class. Like Courtenay he seems to have acquired independent wealth, yet one cannot detect a whisper of his involvement in plunder. He was respected by all and was addressed reverently by the port authorities as m[aste]r Newton, distinct from other captains. However his ships, on the high seas since 1564, were armed for his own defence.[21] Newton had received nothing for his efforts in the Arctic except a cargo of worthless ore (at his own request) and a menial £10. He seems to have been hurt by the loss of his buss. By 1583 his trading had waned and at the then-ripe age of 52, he probably looked forward to impending retirement. We have equated him with Richard Newton of Bridgwater who died May 11, 1587, leaving a small acreage of "tilled land, windmills, orchards, 300 acres of pasture, 100 acres of meadowland and 37 acres of woodland".[22] Just how he acquired this farm is not known.

But for the most of the company, the voyages heralded a brief moment of emergence from obscurity. Some reappeared in Edward Fenton's "troublesome voyage" to Africa and South America in 1582-3 (Taylor 1959). Edward Robinson and Esdras Draper received promotions; John Smythe, Christopher Jackson, Nicholas Chancellor, Peter Robinson and David Evans got their old jobs back. Richard Fairweather, the younger, was master's mate, under Christopher Hall, in the *Galleon Leicester*. For Chancellor and Evans it was to be their last voyage: they died mid-1582. Fairweather left the Company at Rio Plate, where he married and took up permanent residence.

Frobisher and Lok, the principal players

Frobisher and Lok were contrasting characters: Frobisher with his maritime upbringing and Lok with his start in the mercantile business; Frobisher the impetuous and violent and Lok the methodical and calm; Frobisher almost illiterate and Lok multilingual and articulate; Frobisher the escapee (in the northwest disaster) and Lok the scapegoat.

Frobisher's early years are known for acts of piracy and dubious exploration, dating from his teens in the Guinea voyages of 1553 and 1554. These led to other voyages of gain and plunder (Marsden 1906; Eliot 1917). His exploits must have met with the approval of the Queen because, about 1570, she made him Clerk of the Ships in reversion (upon decease or default of 'G.W.') "in consideracion of the true and faithfull service hertofore donne".[23] In 1572 he joined Englands maritime force, thus setting a trend which lasted his lifetime. After the northwest venture, Frobisher (the famous) was never long unemployed. In the winter of 1579 he was charged with transporting Duke John Casimir with mariners and guard of 110 in the *Foresight* to Flushing.[24] The following year he was once more in the *Foresight*, this time to participate in the siege of Smerwick but, as noted in Chapter 3, he did not appear. His later distinguished military career is well known (Jones 1578; McFee 1928) but the fact that, after the voyages, he returned to piracy on several occasions[25] has been overlooked. Nevertheless, after 1578, he became a comparatively respectable citizen: his part in the voyages had cemented a connection with the Court and there was no urgency to acquire supplementary income. He was killed in action in 1594.

Throughout the numerous documents and publications, Frobisher comes across as an uneducated, rugged, ill-tempered (at times violent) individual. He made decisions quickly (though not always correctly); he was obstinate and categorical, almost to the extent of being deaf to the advice of his confrères. Whereas such accusations in the *doynges* andd *abuses*[26] are, no doubt, greatly exaggerated, the negative facets of his character are corroborated elsewhere. Williamson (1914) has also inferred dishonesty: Frobisher deliberately failed to publicly disclose that the ore mined in the second voyage had not come from the original locality, therefore deceiving the venturers and leaving the way open for a third and very large expedition. However, just when the new location became common knowledge, is a question that cannot be answered. Furthermore, it was up to Lok, not Frobisher, to make the initial announcement. On the other hand he remained steadfast to his purpose (e.g. in complying with his main instructions for the second and third voyages, almost to the letter), he commanded a tight ship, and was loyal and sympathetic to the plight of his mariners.

Michael Lok was a very different person. Born into the large family of the wealthy merchant Sir John Lok, he remained in school until the age of 13, after which he travelled extensively on family business in Europe and acquired a knowledge of French, Italian, Spanish and Latin. Being Protestant, he remained on the continent during Queen Mary's reign (1553-58) and returned to England only after the accession of Elizabeth. In 1571 he was appointed London agent of the Muscovy Company, a prestigous and lucrative position. Lok seems to have been committed to orthodox trade but "he longed for pre-eminance among his peers".[27] This ambition contributed in a decision to wager his entire fortune in the precarious quest for the Northwest Passage and the unproved mineral wealth of Meta Incognita. It was Lok who invented the enterprise, Lok who brought Frobisher into the picture, and Lok who was the largest private backer. But in the end, it was Lok who was held responsible for the entire debt of the Company. In 1578 his fortunes plummeted, never to recover. By 1581 we find him in debtor's prison but he was at liberty in 1582 and in 1615, at the age of 83, he was still pursued by one creditor. Richard Madox, in 1582, summed up his character and career as follows: "Lok however, is a man of great wit and admirable honesty, as the master [Fenton] reported to me, but unhappy"[28].

Now our story has ended and we can reflect the benefits and troubles resulting from the northwest venture. For the vast majority of the ships' company and its associates, we have little information. Probably most of the mariners did not suffer greatly and some may have prospered: they were in the public eye, heros of the day. For some, the principals, we can hazard general conclusions: on the basis of their experience in Baffin, James Aldaye, Edward Fenton and Charles Jackman were awarded commissions in new voyages of exploration; Christopher Hall and Edward Kendall became renouned pilots; Thomas Courtenay, Henry Moyle, John Smithe, Luke Ward and Gilbert York rendered valuable service to the navy; and Robert Denham managed mines in Cornwall and Wales. It was through revitalized support of the Court that Martin Frobisher was given important naval assignments, and support of the Borough that Hugh Randall received new civic appointments. Nicholas Dennys and Richard Newton lost their ships, suffered immediate hardships, but later overcame these misfortunes. The ventures that had difficulty in meeting their pledges simply did not pay up, speculators probably became more suspicious of dubious mining developments, but they were now aware of potential mineral resources in the far north.

Of the key personnel, only Michael Lok, who was held responsible for the outstanding pledges, and Jonas Shutz, who left his furnaces in ill health and disgrace, were drastically afflicted. As for the Cathay Company, failure was complete. The low (but realistic) assay values, although omitted from official documents, appear to have become common knowledge by 1578. The public lost heart and the third voyage lacked the tumultous homecoming of the second.   One may question the judgement of the commissioners for having encouraged so transparent an enterprise. Even if the ore had lived up to the high initial expectations (that were repeatedly questioned), there still remained considerations of leadership (which were certainly wanting) and the Dartford plant (which was poorly constructed). Obviously, the project was doomed from the start!

These effects must be regarded as short term. In the long run the enterprise ushered in an era of European awakening in northern geography. In particular, the voyages were the first of many firmly established British sovereignty over the Canadian Arctic islands.

Notes

1. Notable is a map by van der Aa (1729, 20: plate 24.29) showing a hypothetical route taken in the first voyage along with an imaginary drawing of the *Gabriel, Michael* and a pinnace. The error in location of Frobisher Straits, and efforts of Hans Egede to locate it, are told by Kejlbo (1971).

2. On the map are five arrows of magnetic variation, the first map of its kind. The arrows indicate not only the regional compass variation but also the trajectory of Frobisher's first voyage (Christy 1900: 47-49; *SM* 1: ff xcic; Skelton *et al.* 1971: 69 and map 6).

3. *BL*, Cotton, Otho E8, No. 16, fol. 78, 1580. The map has been discussed by Taylor (1938).

4. This compilation was made from manuscript maps, now lost, in about 1636. It contains details not found elsewhere such as ice choked bays (one of which terminates 'Frobisher Straits'!), anchorages and ocean depths.   The manuscript map is in the Bayerische Staatsbibliothek, Munich. It may have been drawn on information supplied by Captain Abraham Kendall.

5.   This map, probably engraved by Francis Gualle, was republished in facsimile by Nordenskiold (1889: map 131) and reproduced by Parks & Williamson (1961: Fig. 20).

6. Meta Incognita [Baffin Island] was described on parchment as "the islande, and broken land easterly, and somewhat to the south of Labrador [Greenland], [which] were particularly discovered and possessed A[nn]o. 1576. and the last yere [1577] by Martin Frobysher Esquier". *BL*, Cotton, Augustus 1, No. 1.

7. The story of Frobisher's Inuit is described in detail by Cheshire *et al.* (1980). Further details can be found in Settle and the chronicle of William Adams (*BRO*, MS 13748/4, 1576-1577).

8. Information was available, for example, in woodcuts and descriptions of natives, their kayaks and whale-bone huts in Olaus Magnus (1555: 69-70, 621, 754).

9. *PRO*, E 164/35: 30, 128. "A greate picture of the whole bodye of the strainge man in his garments ... geven the Quenes ma[jes]tie. Another like picture for the Companye. ij smalle pictures of his head [one for the Queen, one for the company]. A moulde of hard earth for the tartar man ymage". Total cost to the company of paintings and frames was £14 14s 8d.

10. *PRO*, E 164/35: 150. "A greate picture of the strainge womane for the Quene. A small picture of the strainge womane for the Quene. A greate picture of the strainge womane for the Companye. A small picture of the strainge womane for the Companye". Total cost to the company of paintings, frames and cartage was £11 4s 0d.

11. Besides the 8 paintings by Ketel and a cast by Cure (none of which are known to have survived), the following pictures of Frobisher's Inuit are on record: wash by Lucas de Heer (in University Library of Ghent), engraving by Marcus Gheeraets (in Rijksmuseum) and 5 water colour portraits in the John White album (British Museum), see Hulton (1961), Hulton & Quinn (1964), Hulton (1984). Cheshire *et al.* (1980) also mention sketches by Adriaen Coenenzn, made in late 1577 or early 1578, copied from pictures of an unidentified mariner in the Hague.

12. Settle (1578, 1580), Anania (1582).

13. Little is known of Robert Denham's early or private life. A family of Denhams lived on Matthew Friday Street, London (Society of Genealogists, Percival Boyd MS *c*. 1930, *Citizens of London*) including William Denham, goldsmith (died Aug. 31, 1583) and Thomas Denham, goldsmith (died April 11, 1586, aged 64 years). James, son of Robert Denham, goldsmith, and Elizabeth Denham, was baptized at St Olav's Church, Hart Street (not far from Matthew Friday Street) in 1565. An unidentified Denham, goldsmith of Cheapside, was involved in the plot to marry Mary Queen of Scots to the Duke of Norfolk (Testimony of Robert Higford, October 21, 1569. *CP*, MS 5, Nos. 5-9). Robert Denham appears to have died on or before 1605 (Admin. Elizabeth Denham, widow, St. Bot., Alders., Reg. 4, 290).

14. Edward Fenton visited this locality in late December 1578 and collected ore from the spoils, on the property of John Nance within 2 miles [3km] of Newquay ['New Kaie'], and the lode, 1 foot [0.3 m] broad, then stood below 22 fathoms [40 m] of water. Fenton also visited the lead-silver lode at nearby St. Columb Minor, worked by Burchard Kranich and later by Robert Denham. He described it as "hard by the see side, the loade scant a foate [foot] broade". Fenton to Privy Council, January 2 1579 [NS]: *PRO*, SP 12/129/2 [*SM* 2: 149].

15. The mines in Cornwall, Devon and Cardigan and mining privileges were leased by Smythe for £300 per annum. Royalties to the Queen were paid by the parent company as a portion of all silver (10%) and copper (5%) contained in the ores (Pettus 1670: 50-1; *BL*, Lansdowne MS 47/66 quoted by Scott 1968; 2: 395-6). Incoming correspondence (42 letters) to the partners is preserved in the Public Record Office (SP 12/163/46 et seq.). Twelve documents are transcribed in full in Grant-Francis (1881). Another source for this paragraph is the William Carnsew MS, c. 1586: *CRO*, DDME (821).

16. This quotation is from: Anon. Spanish advertisements [intelligence], June 23, 1580, *PRO*, SP 94/1/51; Cal. SP, Foreign Ser., Eliz. I 14, No. 336: 314-5.

Other information in this paragraph has been compiled from the following sources:

1. Don Luis de Requesens, Letters of Marque to William Cotton and Henry Carew [Henrrico Careu], May 18, 1575, *PRO*, SP 70/134/118; Cal. SP, Foreign Ser., Eliz. I **10**, No. 132: 57. [Contemporary copy, in Spanish].
2. H. Cobham to F. Walsingham, Jan. 13, 1580, *PRO*, SP 78/4(1)/4; Cal. SP, Foreign Ser., Eliz. I **14**, No. 132: 129.
3. Anon. Advertisements [intelligence] from France, April 27, 1580, *PRO*, SP 78/4(1)/63; Cal. SP, Foreign Ser., Eliz. I **14**, No. 279: 250.
4. H. Cobham to F. Walsingham, June 15, 1580, *PRO*, SP 78/4(1)/93; Cal. SP, Foreign Ser., Eliz. I **14**, No. 327: 309.

'Captain Carew' garrisoned at Holy Island and Berwick-upon-Tweed, north England, in 1560-1563 (*PRO*, Cal. SP, Foreign Ser., Eliz. I **3 - 6**), may be a different person.

17. The biographical sketch is mainly taken from Lloyd (1967) and from the Sherren Papers in the Weymouth Borough Archives (MSS S32- 162). Pertinent data are also found in *PRO*, SP 12/113/9, 24; SP 12/123/37; SP 12/135; HCA 1/35/51-2; HCA 1/36/361-2; HCA 1/37/89- 91; HCA 1/38/5, 70.

18. *PRO*, SP 63/69/68 (Oct. 17, 1579); SP 63/70/38 (Nov. 29, 1579). The quotation is from the latter document.

19. Patent Rolls, Eliz. 1, Pt 3 (Feb. 16, 1580). *PRO*, C 66/1188, m. 5.
The quotation is a translation from the Latin.

20. *PRO*, Court of the Star Chamber; Stac 5, P 64125 (1585). Ellis Pomeroy and Walter Marche vs. Phillip Harte, William Courtenay, James Courtenay and Thomas Courtenay.

21. Water Bailiffs Accounts; *SRO*, D/B/bw 1469 (1572), 1470 (1574), 1471 (1582).

22. Inquisitions post mortem; *PRO*, C 142/277/29; Richard Newton (d. May 11, 1587). The quotation is a translation from the Latin. The document is a parchment strip which is partly illegible and partly missing due to rodent gnawing.

23. Letters patent from the Court of Chancery to Martin Frobisher ['M.F']. *PRO*, SP 40/1 f. 119 [n.d.].

24. Annual report of Sir John Hawkins, treasurer of the navy, for 1579. *PRO*, E 351/2215. Frobisher was appointed captain and George Beeston ['Beston'] master.

25. *SRO*, Phelips Family Papers (1590). "Articles Exhibited by Monsr Caron in the name of the Estates General". 1, in June and July 1590 Sir Martin Frobisher and Sir John Hawkins took [from Dutch vessels] a great quantity of "Cochenill [cochineal], packs and flanels". 2, Young Martin Frobisher [son of Davy, Sir Martin's brother], under 'conduct' of Sir Martin, took from the ship of Peter Peterson 58 bags of money, cochineal and other goods.

Frobisher also took part in an action (1593) sponsored by Sir Thomas Myddelton and directed by Sir Walter Raleigh. He appears to have captained the bark *Disdain* in a mission of reprisal against the Spanish and sold prizes in Dartmouth. See *NLW*, Chirk Castle MS, F12540, pp. 139-40 (1593-1594).

26. *The abuses of Captayn Furbusher* summarize *The doynges of Captayne Furbusher* (discussed in Chapter 1). The former are transcribed in *SM* 2: 208-12.

27. This quotation is taken, and the paragraph summarized, from McDermott (1984: 1-22).

28. Quotation from the diary of Richard Madox (transcribed in Donno 1976: 224).

REFERENCES

Aa, P.v.d. 1729. *La galerie agréable du monde*. Chez Pierre van der Aa, Leide, 27 vols.

Agricola, G. 1561. *De re metallica*. Second Latin edition. Froben, Basle.

Ala, D. 1992. *A petrochemical study of ultramafic rocks from the Countess of Sussex Mine, Baffin Island, N.W.T.* B.Sc. thesis, University of Ottawa. 60 pp.

Alsford, S. [ed.] 1993. The Meta Incognita Project: contributions to field studies. *Canadian Museum of Civilization, Mercury Series*, Directorate Paper **6**.

Anania, G.L. 1582. *Lo Scoprimento dello Stretto Artico et di Meta Incognita retrovato nel'anno MDLXXVII & 1578 dal Capitano Martino Forbisero Inglese*. Gio. Battista Cappelli, Napoli.

Anderson, R.C. [Compiler] 1959. List of English men-of-war. *Society for Nautical Research* Occasional Publication **7**.

Andrews, K.R. [ed.] 1959. *English privateering voyages to the West Indies 1588-1595*. Cambridge (Hakluyt Society, Ser. 2, No.111).

-------- [ed.] 1972. *The last voyage of Drake & Hawkins*. Cambridge (Hakluyt Society, Ser. 2, No.142).

Anspach, Rev. L.A. 1819. *A history of the island of Newfoundland, containing a description of the island, the Banks, the fisheries, and trade of Newfoundland and the coast of Labrador*. London (published by the author).

Babcock, W.H. 1922. *Legendary islands of the Atlantic*. American Geographical Society, New York.

Bagwell, R. 1890. *Ireland under the Tudors*. Longmans, Green, & Co., London. 3 vols.

Barnes, S.-J. & Duke, J.M. [eds.] 1990. Advances in the study of platinum - group elements. *Canadian Mineralogist* **28**: 377-689.

Barrow, J. 1818. *A chronological history of voyages into the Arctic regions*. John Murray, London.

Baughman, R. 1938. Variant editions of Settle's account of Frobisher. *The Huntington Library Quarterly* **2**: 67-8.

Becher, A.B. 1833. Extract from records of the King's Rememberancer's Office at the King's Mews:- Sir Martin Frobisher's Voyages for the Discovery of a North-West Passage. *Nautical Magazine* **2**: 470-6.

-------- 1843. The voyages of Martin Frobisher. *Journal of the Royal Geographical Society of London* **12**: 1-20.

Berger, G.W. & York, D. 1981. Geothermometry from ^{40}Ar/^{39}Ar dating experiments. *Geochimica et Cosmochimica Acta* **45**: 795-811.

Best, G. 1578. *A true discourse of the late voyages of discoverie, for the finding of a passage to Cathaya by the Northweast, under the conduct of Martin Frobisher, Generall.* Henry, Bynnyman, London.

------- 1900. *Frobisher's first voyage from "a true discourse"* [Copy of part of the 1578 edition]. Old South Leaflets, [Boston] General Series **5** (117).

Biggar, H.P. 1932. The naming of America and Greenland. *Canadian Geographical Journal* **4**: 85-96.

Blackadar, R.G. 1967a. Geological reconnaissance, southern Baffin Island, District of Franklin. *Geological Survey of Canada* Paper **66-47**.

-------- 1967b. Kodlunarn Island and Frobisher's gold. *The Arctic Circular* **17**: 1-12.

Boetzkes, O. 1964. *Sir Martin Frobisher's search for the northwest passage.* Exposition Press, New York.

Boreham, P.W. 1986a. The history of Dartford Priory and King Henry VIII's manor house at Dartford. *Dartford Local History Leaflet* **13**.

-------- 1986b. Tudor Dartford. *Dartford Local History Leaflet* **14**.

-------- 1991. *Dartford's royal Manor House re-discovered.* Dartford Borough Council, Dartford.

Bourne, H.R.F. 1868. *English seamen under the Tudors.* Richard Bentley, London.

Boyle, R.W. 1979. The geochemistry of gold and its deposits (together with a chapter on geochemical prospecting for the element). *Geological Survey of Canada* Bulletin **280**.

------- 1982. Gold, silver, and platinum metal deposits in the Canadian Cordillera - their geological and geochemical setting. In *Precious Metals in the Northern Cordillera*. Association of Exploration Geochemists: 1-19.

Bruemmer, F. 1966. Kodlunarn Island's gold rush. *Canadian Geographical Journal* **72**: 49-51.

Brugerolles, E., Bari, E., Benoît, P., Fluck, P. & Schoen, H. [eds.] 1992. *La mine mode d'emploi. La rouge myne de Sainct Nicolas de la Croix, dessinée par Heinrick Groff.* Gallimard, Paris.

Buerger, M.J. 1938. Spectacular Frobisher's Bay. *Canadian Geographical Journal* **17**: 3-17.

Byrne, M.J. 1903 [translator]. *Ireland under Elizabeth. A portion of the history of Catholic Ireland by Don Philip O'Sulivan Bear, translated from the original Latin.* Sealy, Bryers & Walker, Dublin.

Caswell, J.E. 1969. The sponsors of Canadian arctic exploration. *The Beaver* **Spring 1969**: 4-13.

Cheshire, N., Waldron, T., Quinn, A. & Quinn, D. 1980. Frobisher's Esquimos in England. *Archivaria* **10**: 23-50.

Christy, M. 1897. On 'Buss Island'. In Gosch, C.C.H. [ed.] *Danish arctic exploration 1605-1620.* London (Hakluyt Society, ser. 1, No.96): 164-202.

------- 1900. *The silver map of the world, a contemporary medallion commemorative of Drake's great voyages (1577-80).* Henry Stevens, Son & Stiles, London.

Cleaveland, E. 1735. *A genealogical history of the noble and illustrious family of Courtenay.* Edward Farley, Exon.

Cochran-Patrick, R.W. 1878. *Early records relating to mining in Scotland.* David Douglas, Edinburgh.

Collingwood, W.G. 1912. *Elizabethan Keswick.* Titus Wilson for the Cumberland and Westmoreland Antiquarian and Archaeological Society, Kendal.

Collinson, R. (ed.) 1867. The three voyages of Martin Frobisher in search of a passage to Cathaia and India by the north-west. London (Hakluyt Society, Ser. 1, No. 38).

Cooke, A. 1964. Canada's first gold rush. *The Beaver* **Summer 1964**: 24-7.

Cooper, C.P. [ed.] 1833. Record's of King's Remembrancer's office at the King's Mews. Sir Martin Frobisher's voyages for the discovery of a north west passage. *Proceedings of H.M. Commissioners on Public Records* **1832-1833**: 74-7, 558-62.

Corbett, J.S. 1898a. Papers relating to the Navy during the Spanish War 1585-1587. *Naval Record Society* **11**, London.

------- 1898b. *Drake and the Tudor navy.* Longmans, Green, and Co., London. 2 vols.

Cuppage, J., Bennett, I., Cotter, C. & O Rahilly, C. 1986. *Archaeological survey of the Dingle Peninsula.* Ballyferriter (Oidhreacht Chorca Dhuibhne), Dublin.

Dalrymple, A. 1789. *Memoir of a map of lands around the north pole.* George Bigg, London (Replica Copy, 1973. I. Ehrlich, Montreal).

Dalrymple, G.G. & Lanphere, M.A. 1969. *Potassium - Argon dating.* W.H. Freeman & Co., San Fransisco.

Dartford District Archaeological Group 1986. *Rediscovering Dartford.* Dartford.

Donald, M.B. 1950. Burchard Kranich (c.1515-1578), miner and Queen's physician, Cornish mining stamps, antimony, and Frobisher's gold. *Annals of Science* **6**: 308-322.

------- 1951. A further note on Burchard Kranich. *Annals of Science* **7**: 107-8.

------- 1955. *Elizabethan copper: The history of the Company of Mines Royal, 1568-1605.* Pergamon, London.

------- 1961. *Elizabethan monopolies: the history of the Company of Mineral and Battery Works.* Oliver and Boyd, Edinburgh.

Donno, E.S. [ed.] 1976. *An Elizabethan in 1582. The diary of Richard Madox, Fellow of All Souls.* London (Hakluyt Society, Ser. 2, No. 147).

Ehrenreich, R.M. 1993. An evaluation of the Frobisher iron blooms: a cautionary tale. In Fitzhugh, W.W. & Olin, J.S. [eds.] *Archeology of the Frobisher voyages.* Smithsonian Institution Press, Washington, DC: 221-8.

Eliot, K.M. 1917. The first voyages of Martin Frobisher. *English Historical Review* **32**: 89-92.

Ellis, H. 1816. Sir Martin Frobisher's instructions when going on a voyage to the north-west parts and Cathay in the time of Queen Elizabeth. *Archaeologia, or Miscellaneous Tracts relating to Antiquity* **18**: 287-90.

Ellis, T. 1578. *A true report of the third and last voyage into Meta incognita.* Thomas Dawson, London.

------- 1922. *A true report* ... [facsimile of the 1578 edition, above] Massachusetts Historical Society, Boston.

Fitzgerald, R.O. 1855. Fort-del-Ore, A.D. 1580, [reconstructed and] reduced from the original plan in the State Paper Office for the Kerry Magazine by R.O. Fitzgerald. *Kerry Magazine* **2**: 80a.

Fitzhugh, W.W. 1993a. Archeology of Kodlunarn Island . In Fitzhugh, W.W. & Olin, J.S. [eds.] *Archeology of the Frobisher voyages.* Smithsonian Institution Press, Washington, DC: 56-97.

-------- 1993b. Field surveys in outer Frobisher Bay. In Fitzhugh, W.W. & Olin, J.S. [eds.] *Archeology of the Frobisher voyages.* Smithsonian Institution Press, Washington, DC: 98-135.

-------- 1993c. Questions remain. In Fitzhugh, W.W. & Olin, J.S. [eds.] *Archeology of the Frobisher voyages.* Smithsonian Institution Press, Washington, DC: 229-238.

-------- & **Olin, J.S.** [eds.] 1993. *Archeology of the Frobisher voyages.* Smithsonian Institution Press, Washington, D.C.

Francis, J.P. 1969. Robert Wolfall - first Anglican clergyman in Canada. *Journal of the Canadian Church Historical Society* **11**: 2-32.

Gad, F. 1971-82. *The history of Greenland.* McGill-Queen's University Press, Montreal.   3 vols.

Glasgow, T. & Salisbury, W. 1966. Elizabethan ships pictured on Smerwick map, 1580. Background, authentication and evaluation. *Mariner's Mirror* **52**: 157-65.

Gonciar, A. 1993. *A petrochemical study of ultramafic rocks from Tikkoon Peninsula, Baffin Island, N.W.T.* B.Sc. thesis, University of Ottawa. 83 pp.

Gowan, M. 1979. *Irish artillery fortifications 1550-1700.* M.A. thesis, University College, Cork.

Grant - Francis, Col. G. 1881. *The smelting of copper in the Swansea district of south Wales from the time of Elizabeth to the present day.* Henry Sotheran & Co., London and Manchester.

Gunther, R.T. 1927. The great astrolabe and other scientific instruments of Humphrey Cole. *Archaeologica, or Miscellaneous Tracts Relating to Antiquity* **76**: 273-317.

Hakluyt, R. 1587. *De orbe novo Petri Martyris Anglerii ... decades octo ... labore & industria Richardi Hakluyti.* Guillaume Auuray, Paris.

------ 1589. *The principall navigations, voiages and discoveries of the English nation.* George Bishop & Ralph Newberrie, London.

------ 1598-1600. *The principall navigations, voyages, traffiques and discoveries of the English nation.* George Bishop, Ralph Newberrie & Robert Barker, London. 3 vols.

------ 1927. *The principal voyages, traffiques and discoveries of the English nation.* Third edition. J.M. Dent & Sons, London. 8 vols.

Hall, C.F. 1864. *Life with the Esquimaux.* Sampson Low, Son and Marston, London. 2 vols.

Hammersley, G. [ed.] 1988. *Daniel Hechstetter, the younger. Memorabilia and letters, 1600-1639. Copper works and life in Cumbria.* Franz Steiner, Stuttgart.

Harbotte, G., Cresswell, R.G. & Stoenner, R.W. 1993. Carbon - 14 dating of iron blooms from Kodlunarn Island. In Fitzhugh, W.W. & Olin, J.S. [eds.] *Archeology of the Frobisher voyages.* Smithsonian Institution Press, Washington, DC: 172-80.

Harbottle, G. & Stoenner, R.W. 1987. Carbon 14 dating of iron blooms from Kodlunarn Island, Baffin Island, Canada. *Society for Historical Archaeology Annual Meeting* [Jan. 7-11 1987, Savannah, Georgia] Program and Abstracts: 58-9.

[Hitchock, R.] 1854. The antiquities of Kerry. No. VII, the true history of Fort-del-Ore, Smerwick Harbour. *The Kerry Magazine* **1** (8): 113-116.

Hogarth, D.D. 1985. Petrology of Martin Frobisher's "black ore" from Frobisher Bay, N.W.T. *GAC-MAC Annual Meeting*, Program with Abstracts **10**: A 28.

------- 1989. The *Emanuel* of Bridgwater and Discovery of Martin Frobisher's "black ore" in Ireland. *The American Neptune* **49**: 14-20.

------- 1990. Field investigation of Martin Frobisher's mines and furnace sites in SE Baffin Island, August 7-12, 1990. Unpublished report to Arctic Studies Program, Smithsonian Institution, Washington, DC.

------- 1993a. Mining and metallurgy of the Frobisher ores. In Fitzhugh, W.W. & Olin, J.S. [eds.] *Archeology of the Frobisher voyages*. Smithsonian Institution Press, Washington, D.C.: 136-145.

------- 1993b. The ships' company in the Frobisher voyages. In Fitzhugh, W.W. & Olin, J.S. [eds.] *Archeology of the Frobisher voyages*. Smithsonian Institution Press, Washington, D.C.: 15-16; 241-251.

------- 1993c. The impermanence of Kodlunarn Island. In Alsford, S. [ed.] The Meta Incognita Project: contributions to field studies. *Canadian Museum of Civilization, Mercury Series*, Directorate Paper **6**: 81-88.

------- **& Gibbins, W.A.** 1984. Martin Frobisher's "gold mines" on Kodlunarn Island and adjacent Baffin Island, Frobisher Bay, NWT. *Contributions to the Geology of the Northwest Territories* **1**: 69-78.

------- **& Loop, J.** 1986. Precious metals in Martin Frobisher's 'Black ores' from Frobisher Bay, Northwest Territories. *Canadian Mineralogist* **24**: 259-63.

------- **Loop, J. & Gibbins, W.A.** 1985. Frobisher's gold on Kodlunarn Island - fact or fable? *CIM Bulletin* **78**: 75-9.

-------, **Moore, D.T. & Boreham, P.W.** 1993. Martin Frobisher mines and ores. In Alsford, S. [ed.] The Meta Incognita Project: contributions to field studies. *Canadian Museum of Civilization, Mercury Series*, Directorate Paper **6**: 148-75.

------- **& Roddick, J.C.** 1989. Discovery of Martin Frobisher's Baffin Island "ore" in Ireland. *Canadian Journal of Earth Science* **26**: 1053-60.

Holinshed, R. 1587. *The second volume of the chronicles: conteining the description, conquest, inhabitation and troublesome estate of Ireland. First collected by Raphaell*

Holinshed ... augmented by John Hooker alias Vowell gent. Part 2. The Irish History.
John Harrison, George Bishop, Ralph Newberie, Henry Denham & Thomas Woodcock,
London.

Hoover, H.C. & Hoover, L.H. [translators & eds.] 1950. *De re Metallica.* Dover
Publications, New York.

Horne, R.R. 1974. The lithostratigraphy of the late-Silurian to early Carboniferous of
the Dingle Peninsula, Co. Kerry. *Geological Survey of Ireland*, Bulletin **1**: 395-428.

-------- 1976. Geological guide to the Dingle Peninsula. *Geological Survey of Ireland*
Guide series 1.

-------- **& MacIntyre, R.M.** 1975. Apparent age and significance of Tertiary dykes in
the Dingle Peninsula, southwest Ireland. *Scientific Proceedings of the Royal Dublin
Society*, Series A **5**: 293-99.

Hulton, P.H. 1961. John White's drawings of Esquimos. *The Beaver* **Summer 1961**:
16-20.

-------- 1984. *America 1585: the complete drawings of John White.* The University of
North Carolina Press, Raleigh, and British Museum Publications, London.

-------- **& Quinn, D.B.** 1964. *The American drawings of John White. 1577-1590.*
British Museum, London. 2 vols.

Hume, M.A.S. (ed.) 1894. *Letters and state papers relating to English affairs,
preserved in the Archives of Simancas* **2** (**Elizabeth I, 1568-1579**). British National
Archives, London.

Jackson, D.D. 1993. Hot on the cold trail left by Sir Martin Frobisher. *Smithsonian*
23 (10): 119-130.

Jackson, G.D. & Morgan, W.C. 1978. Precambrian metamorphism on Baffin and
Bylot Islands. *Geological Survey of Canada*, Paper **78-10**: 249-67.

Johnson, A.M. 1942. The mythical land of Buss. *The Beaver* **Outfit 273** (December
1942): 43-47).

Jones, F. 1878. *The life of Sir Martin Frobisher, knight, containing a narrative of the
Spanish Armada.* Longmans, Green & Co., London.

Jones, F.M. 1954. The plan of the golden fort at Smerwick, 1580. *The Irish Sword* **2**: 41-2.

Kamineni, D.C., Jackson, G.D. & Bonardi, M. 1979. Coexisting magnesian and calcic amphiboles in meta-ultramafites from Baffin Island (Arctic Canada). *Neues Jahrbuch für Mineralogie, Monatschrift* **1979** (**12**): 542-55.

Kejlbo, I.R. 1971. Hans Egede and the Frobisher Strait. *Geografisk Tidsskrift* **70**: 59-105.

Kenyon, W.A. 1975a. All is not golde that shineth. *The Beaver* **Summer 1975**: 40-6.

------- 1975b. *Tokens of possession: the northern voyages of Martin Frobisher.* Royal Ontario Museum, Toronto.

------- 1980. The Reverend Robert Wolfall in Arctic Canada, 'A watry pilgrimage' of 1578. *Rotunda* **13**: 7-11.

------- [ed.] 1981. The Canadian arctic journal of Capt. Edward Fenton 1578. *Archivaria* **11**: 171-203.

King, F.A. 1955. When Martin Frobisher searched for gold. *Canadian Mining Journal* **76**: 67-9.

Kirkham, N. 1969. Early lead smelting in Derbyshire. *Transactions of the Newcomen Society* **41**: 119-38.

Klarwill, V. von [translator and ed.] 1924. *The Fugger news letters. Being a selection of unpublished letters from the correspondents of the House of Fugger.* John Lane, London.

Kretz, R. 1982. Transfer and exchange equilibria in a portion of the pyroxene quadrilateral as deduced from natural and experimental data. *Geochimica et Cosmochimica Acta* **46**: 411-421.

Laughton, J.K. [ed.] 1895. *State papers relating to the defeat of the Spanish Armada anno 1588.* Naval Records Society, London. 2 vols.

Leslie, Sir. J., Jameson, R. & Murray, H. 1855. *The polar seas and regions.* Nelson & Sons, London.

Li, T. & Yio, C. 1966. The abundance of chemical elements in the earth's crust and its major tectonic units. *Scientia Sinica* **15**: 258-72.

Lloyd, R. 1967. *Dorset Elizabethans at home and abroad.* John Murray, London.

Lodge, E. 1827. *Life of Sir Julius Caesar with memoirs of his family and descendants.* John Hatchard & Son, London.

Lonicer, A. 1551. *Naturalis historiae opus novum.* Apud Chr. Engenolphum, Francofurti.

Manhart, G.B. 1924. The English search for a north-west passage in the time of Queen Elizabeth. In, Rowland, A.L. & Manhart, G.B. (eds). *Studies in English commerce and exploration in the reign of Elizabeth* **2**: 1-179 University of Philadelphia, Philadelphia.

Marsden, R.G. 1906. The early career of Sir Martin Frobisher. *English Historical Review* **21**: 538-44.

Martin, W. 1983. Once upon a mine: story of pre-confederation mines on the island of Newfoundland. *Canadian Institute of Mining and Metallurgy*, Special Volume **26**.

McDermott, J. 1984. *The account books of Michael Lok, relating to the northwest voyages of Martin Frobisher 1576-1578: text and analysis.* M. Phil. thesis, University of Hull. 512 pp.

McFee, W. 1928. *The life of Sir Martin Frobisher.* Harper & Brothers, New York.

McGhee, R. & Tuck, J.A. 1993a. An archaeological assessment of Qallunaaq Island. In Alsford, S. [ed.] The Meta Incognita Project: contributions to field studies. *Canadian Museum of Civilization, Mercury Series*, Directorate Paper **6**: 7-27.

------- & ------- 1993b. An Elizabethan settlement in Arctic Canada. *Rotunda* **25**(4): 32-40. Letters, *ibid.* **26**(1): 45; **26**(2): 47.

Moore, D.T. & Oddy, W.A. 1985. Touchstones: some aspects of their nomenclature, petrography and provenance. *Journal of Archaeological Science* **12**: 59-80.

Moray, Sir R. 1665. A way to break easily and speedily the hardest rocks, communicated by the same person, as he received it from Monsieur Du Son, the inventor. *Philosophical Transactions* **1**: 82-85.

Nordenskiöld, N.A.E. 1889. *Facsimile-atlas to the early history of cartography.* P.A. Norstedt & Sons, Stockholm.

Olaus Magnus 1555. *Historia de gentibus septentrionalibus ....* Vatican Press, Rome [Reissued 1972 with introduction by John Granlund. Rosenkilde & Bagger, Copenhagen].

Olin, J.S. 1993. History of research on the Smithsonian Bloom. In Fitzhugh, W.W. & Olin, J.S. [eds.] *Archeology of the Frobisher voyages.* Smithsonian Institution Press, Washington, D.C.: 48-55.

Oppenheim, M. [ed.] 1902. The naval tracts of Sir William Monson in six books. *The Naval Record Society* **22, 23, 43, 45, 47.**

O'Rahilly, A. 1937. The massacre at Smerwick (1580). *Journal of the Cork Historical and Archaeological Society* **42**: 1 - 15; 65 - 83.

Parks, G.B. 1935. Frobisher's third voyage, 1578. *Huntingdon Library Bulletin* **7**: 181-90.

------- 1938. The two versions of Settle's Frobisher narrative. *The Huntingdon Library Quarterly* **2**: 59-65.

------- **& Williamson, J.A.** 1961 (2nd edn.). *Richard Hakluyt and the English voyages.* Frederick Ungar, New York.

Pears, S.A. 1845. *Correspondence of Sir Philip Sidney and Hubert Languet.* William Pickering, London.

Pearson, B. 1966. Cathay revisited. *North* **13(3)**: 1-11.

Percy, J. 1880. *Metallurgy: the art of extracting metals from their ores* **3 (1)**, *Silver and Gold.* John Murray, London.

Pettus, Sir J. 1670. *Fodinae regales, or the history, laws and places of the chief mines and mineral works in England, Wales and the English Pale in Ireland ...* Printed by H.L. and R.B. for Thomas Basset, London.

Philp, B. 1986. *The Dartford gunpowder mills.* Kent Archaeological Rescue Unit, Dover.

Pingel, C. 1845. Chapter 33, The most important voyages made in recent time from Denmark and Norway to seek the lost Greenland and bring about its rediscovery. *In* Rafn, C.C. & Magnússon, F. [eds] *Grönlands Historiske Mindesmaerker* **3**: 625-794. Kongelige Nordiske Oldskrift-selskab (Grönlansk Afdeling), Kjøbenhavn. [In Danish].

Porteous, Rev. J.M. 1876. *Gods treasure house in Scotland.* Simpkin, Marshall & Co., London.

Powell, A.H. 1909. *The ancient borough of Bridgwater.* Page & Son, Bridgwater.

Quinn, D.B. 1955. *The Roanoke voyages 1584-1590. Documents to illustrate the English voyages to North America under the patent granted to Walter Raleigh in 1584.* London (Hakluyt Society, Ser. 2, No. 104). 2 vols.

-------- **Quinn, A.M. & Hillier, S.** [eds.] 1979. *New American world, a documentary history of North America to 1612. 4 Newfoundland from fishery to colony. Northwest Passage searches.* Arno Press, and Hector Bye, Inc., New York.

-------- **Skelton, R.A. & Quinn, A.M.** [eds.] 1965. *The principal navigations, voiages and discoveries of the English nation by Richard Hakluyt.* [A photo-lithographic facsimile of the 1589 edition with added introduction and new index] Cambridge (Hakluyt Society, Extra Series, No. 39).

Read, C. [ed.] 1926. Despatches of Castelnau de la Mauvissiere (on Frobisher, Gilbert, de La, Roche, Drake) 1577-81. *American Historical Review* **31**: 285-96.

Redstone, L.J. & members of the London Survey Committee. 1929. The Parish of All Hallows Barking, Part 1. *Survey of London* **12**. London (London County Council).

Richardson, Sir J. 1861. *The polar regions.* Adam & Charles Black, Edinburgh.

Rickard, T.A. 1947. *The romance of mining.* MacMillan, Toronto.

Ritchie, C.I.A. 1964. Dartford's doubtful Eldorado. *Kent Life* **October 1964**: 64-5.

Rodriguez - Salgado, M.J. and the staff of the National Maritime Museum [eds.] 1988. *Armada 1588-1988: an international exhibition to commemorate the Spanish Armada. The Official Catalogue.* Penguin Books, London.

Rowley, S. 1993a. Frobisher miksanut: Inuit accounts of the Frobisher voyages. In Fitzhugh, W.W. & Olin, J.S. [eds.] *Archeology of the Frobisher voyages.* Smithsonian

Institution Press Washington, DC: 26-40.

-------- 1993b. Inuit oral history: the voyages of Sir Martin Frobisher, 1576-78. In Alsford, S. [ed.] The Meta Incognita Project: contributions to field studies. *Canadian Museum of Civilization, Mercury Series*, Directorate Paper **6**: 211-19.

Rowse, A.L. 1969. *Tudor Cornwall: portrait of a society*. MacMillan & Co., London.

Roy, S.K. 1937. The history and petrography of Frobisher's "gold ore". *Field Museum of Natural History. Geological Series* **7**: 21-38.

Rundall, T. 1849. *Narrative of voyages towards the north-west in search of a passage to Cathay in India 1496-1631*. London (Hakluyt Society, Ser. 1, No 5).

Sayre, E.V., Harbottle, G., Stoenner, R.W. Washburn, W., Olin, J.S. & Fitzhugh, W. 1982. The carbon 14 dating of an iron bloom associated with the voyages of Sir Martin Frobisher. In Currie, L.A. [ed.] *Nuclear and Chemical Dating Techniques: Interpeting the Environment Record*. American Chemical Society; ACS Symposium Series **176**: 441-51.

Scott, W.R. 1968. *The constitution and finance of English, Scottish and Irish joint-stock companies to 1720, 2 Companies for foreign trade, colonization, fishing and mining*. Peter Smith, Gloucester, Mass.

Seemann, U. 1984. Tertiary intrusives on the Atlantic continental margin off southwest Ireland. *Irish Journal of Earth Sciences* **6**: 229-35.

Seifert, F & Schumacher, J.C. 1986. Cordierite - spinel - quartz assemblages: a potential geobarometer. *Bulletin of the Geological Survey of Finland* **58**: 95-108.

Settle, D. 1577. *A true reporte of the laste voyage into the West and Northwest regions, &c. 1577*. Henrie Middleton, London.

-------- 1578. *La Navigation du Capitaine Martin Forbisher Anglois, es regions de west & Nordwest, en l'annee MDLXXVII*. Anthoine Chuppin, Geneva.

-------- 1580. *Beschreibung der schiffart des Haubtmans Martini Forbissher ausz Engelland in die Lender gegen West und Nordtwest im Jar 1577*. Katharinam Gerlachin & Johanns vom Berg erben, Nurnberg.

------- 1868. *A true reporte* ... [copy of the 1577 edition in modern script]. John Carter Brown, Providence.

------- 1969. *A true reporte* ... [Facsimile the 1577 edition]. Da Capo Press, New York.

Shammas, C. 1971. *The Elizabethan gentlemen adventurers and western planting.* Ph.D. thesis. The Johns Hopkins University. 239 pp.

------- 1975. The 'Invisible merchant' and property rights: The misadventures of an Elizabethan joint stock company. *Business History* 17: 95-108.

Shaw, W.T. 1983. *Mining in the Lake Counties.* Dalesman Books, Clapham.

Sisco, A.G. & Smith, C.S. [translators & eds.] 1951. *Lazarus Ercker's treatise on ores and assaying, translated from the German edition of 1580.* University of Chicago Press, Chicago.

Skelton, R.A. Summerson, J. & The Marquess of Salisbury 1971. *A description of maps and architectural drawings in the collection made by William Cecil, First Baron Burghley, now at Hatfield House.* The Roxburghe Club, Oxford.

Snoddy, O. 1972. Dún an Óir. *Journal of the Royal Society of Antiquaries of Ireland* **102**: 247-8.

Stefansson, V. & McCaskill, E. [eds.] 1938. *The three voyages of Martin Frobisher.* The Argonaut Press, London. 2 vols.

Steiger, R.H. & Jager, E. 1977. Subcommittee on geochronology: convention on the use of decay constants in geochronology. *Earth & Planetary Science Letters* **36**: 359-362.

Stow, J. 1598. *A Suruay of London, Contayning the Originall, Antiquity, Increase, Moderne estate, and description of that Citie, written in the year 1598 .... by John Stow, Citizen of London.* John Wolfe, London.

Taylor, E.G.R. 1938. Hudson's Strait and the oblique meridian. *Imago Mundi* **3**: 48-52.

------- 1956. *The haven-finding art. A history of navigation from Odysseus to Captain Cook.* Hollis & Carter, London.

------ [ed.] 1959. *The troublesome voyages of Captain Edward Fenton, 1582-1583.* Cambridge (Hakluyt Society, Ser. 2, No. 113).

Todd, S.P., Williams, B.P.J. & Hancock, P.L. 1988. Lithostratigraphy and structure of the Old Red Sandstone of the northern Dingle Peninsula, Co. Kerry, Southwest Ireland. *Geological Journal* **23**: 107-20.

Tuck, J.A., Pilon, J.A. & McGhee, R.J. 1993. The 1991 investigation of Qallunaaq Island. In Alsford, S. [ed.] The Meta Incognita Project: contributions to field studies. *Canadian Museum of Civilization, Mercury Series*, Directorate Paper **6**: 28-54.

Unglik, H. 1993. Metallurgical study of an iron bloom and associated finds from Kodlunarn Island. In Fitzhugh, W.W. & Olin, J.S. [eds.] *Archeology of the Frobisher voyages.* Smithsonian Institution Press, Washington, DC: 181-212.

Waller, G.F. 1979. *Mary Sidney, Countess of Pembroke, a critical study of her writings and literary milieu.* Salzburg (Elizabethan and Renaissance Studies, No.87).

Wanless, R.K., Stevens, R.D., Lachance, G.R. & Edmonds, C.M. 1968. Age determinations and geological studies, K-Ar isotopic ages, Report 8. *Geological Survey of Canada* Paper **67-2A**.

Ward, B.M. 1926. Martin Frobisher and Dr. John Dee. *The Mariners Mirror* **12**: 453-5.

Warner, G.G. [ed.] 1899. *The voyage of Robert Dudley, afterwards styled Earl of Warwick and Leicester and Duke of Northumberland, to the West Indies, 1594-1595 ...* London (Haykluyt Society, ser. 2, No. 3).

Wayman, M.L. & Ehrenreich, R.M. 1993. Metallurgical study of small iron finds. In Fitzhugh, W.W. & Olin, J.S. [eds.] *Archeology of the Frobisher voyages.* Smithsonian Institution Press, Washington, DC: 213-19.

Webb, A. 1878. *A compendium of the Irish biography.* M.H. Gill & Sons, Dublin.

Williamson, J.A. 1914. Michael Lok. *Blackwoods Magazine* **196**: 58-72.

Appendix 1: Assay Technology

Assays by the Cathay Company involved three stages.

1.   Crucible fusion. This operation was performed in the 'melting furnace'. Gold and silver were made to combine, or alloy, with lead; the other metals in the charge went into the slag. The charge may have been ore, lead*, litharge*, charcoal*, saltpetre*, fluorite, borax and soda. Iron and, following the first test at Dartford, copper, were also added, probably as fluxes (Hoover & Hoover 1950: 401). These metals, along with some of the lead, were obtained as sulphides*, which would almost certainly have necessitated a pre-fusion roast.

The charge was melted in large earthenware crucibles or clay pots* and kept above the melting temperature for at least 45 minutes. Then the molten charge was poured, cooled, and solidified, and the 'lead button'*, containing the gold and silver (along with any copper and zinc), was separated by a sharp blow of the hammer from the slag*, containing the remaining metals in the charge.

2.   Cupellation. This operation was performed in the 'fining furnace'. Here all metals, except gold and silver, were extracted with lead oxide. The lead button, with its gold, silver, copper and zinc, was kept at red heat in an open cupel, exposed to a continuous current of air, until all the lead had oxidized. The cupel was moulded from bone ash* at the furnace site and, during cupellation, all the lead, copper and zinc were absorbed into its porous walls. The metallic globule diminished in size until all the lead had gone. A silver-gold 'bead'* remained at the bottom of the cupel.

3.   Parting. This operation involved separating gold from silver in the bead and at Dartford was probably performed in a separate building from parting and cupellation. The principal method involved dissolving silver in nitric acid* and decanting the silver-bearing solution.   Precipitation was effected either by adding brine and precipitating AgCl, or by depositing elemental silver on copper*; both methods were standard practice in 16th-century Germany. An item of nitric acid and an acid flask for the third great proof at London appears in the Exchequer papers (*PRO*, E 164/35, f. 153): "paid for strong water ii lib [2 lb] and a glasse ... for partinge of gold and silver". Elsewhere the acid could have been produced by reacting alum* with saltpetre*, thus explaining the ingredients "Alom and chayne" of Figure 16.

Burchard Kranich used sulphide parting. The silver-gold bead was fused with stibnite ('antimonye')*, thereby sulphidizing the silver and separating the gold in an immiscible antimony - silver alloy (the 'regulus'). Metallic silver was isolated in a second cupellation; gold was recovered by distillation of antimony in a controlled fusion.

Possibly the time-honoured parting method of 'cementation' (Hoover & Hoover, 1950: 456) was also used. Here the silver-gold bead was packed in salt* and powdered brick*, and baked until all the silver had reacted to form AgCl. The product was crushed and sieved, the brittle, pulverized AgCl being undersize, the coarse flattened gold oversize. Silver chloride would then be cupelled. In the Exchequer papers for expenses at Dartford are two items noting salt, one paid October 6 1578 specifying 8 bushels (*PRO*, E 164/36, f. 182).

It would seem that both 'cementation' and stibnite separations were ill-chosen, because these methods were efficient only when the Au:Ag ratio was high (which was certainly not the case with either the northwest ore or the lead additive). A streak test on a touchstone* would have approximately established this ratio, but the assayers, were apparently reluctant to make this test.

  * Additives ('additaments'), materials or products mentioned in contemporary publications or manuscripts.

Appendix II

Potassium-argon isotopic age data
for 'black ores' and associated rocks

Type	A1		B1							C1	C2
Number	E43/2	S/2[e]	E1/1	S/5[e]	88/4	88/6	88/10	CS/1	E23/E	E44/17	88/12A
Locality	E	I	E	I	I	I	I	B	E	E	I
K_2O (wt %)[a]	0.122 ± 0.001	------	0.431 ± 0.004	------	0.135 ± 0.01	0.200 ± 0.003	0.319 ± 0.005	0.353 ± 0.004	0.133 ± 0.003	0.820 ± 0.004	1.51 ± 0.002
Radiogenic ^{40}Ar(mm^{3}g^{-1})[b]	$(1.193\pm0.010)10^{-2}$	------	$(4.83\pm0.04)10^{-2}$	------	$(1.103\pm0.008)10^{-2}$	$(2.270\pm0.016)10^{-2}$	$(3.31\pm0.02)10^{-2}$	$(3.62\pm0.04)10^{-2}$	$(1.278\pm0.010)10^{-2}$	$(6.71\pm0.05)10^{-2}$	$(1.242\pm0.008)10^{-1}$
% Atmospheric Contamination[c]	10.3	------	4.1	------	7.0	5.8	2.8	4.3	5.7	1.4	3.1
Age (Ma)[d] $\pm1\sigma$	1780±20	1810±29[e]	1935±50	1881±30[e]	1580±15	1950±30	1850±30	1830±30	1760±40	1580±15	1590±25

Type	C3		C4	D1	Dolerite	Skarm	Gneiss				
Number	K/9D[e]	K/8B	E43/1	E23/2	88/11A	K/2	K/1	K/7	K/9A	K/10B	K/10C
Locality	B	B	E	E	I	B	B	B	B	B	B
K_2O (wt %)[a]	------	0.592 ± 0.008	0.911 ± 0.004	0.095 ± 0.002	0.450 ± 0.004	1.625 ± 0.004	0.807 ± 0.005	1.020 ± 0.011	0.955 ± 0.004	1.274 ± 0.001	0.887 ± 0.016
Radiogenic ^{40}Ar(mm^{3}g^{-1})[b]	------	$(6.95\pm0.06)10^{-2}$	$(8.35\pm0.06)10^{-2}$	$(7.27\pm0.05)10^{-3}$	$(5.41\pm0.11)10^{-4}$	$(1.051\pm0.006)10^{-1}$	$(5.33\pm0.05)10^{-2}$	$(6.94\pm0.06)10^{-2}$	$(6.63\pm0.06)10^{-2}$	$(8.92\pm0.08)10^{-2}$	$(6.70\pm0.05)10^{-2}$
% Atmospheric Contamination[c]	------	5.2	1.0	8.3	69.8	1.3	2.0	6.1	5.7	3.1	4.0
Age (Ma)[d] $\pm1\sigma$	1722±41[e]	1990±30	1705±15	1510±35	36.9±0.8	1350±10	1370±15	1400±15	1415±15	1425±15	1500±30

[a] Potassium analyses mean of three determinations (flame photometry, Li internal standard).
[b] Radiogenic ^{40}Ar mean of two determinations (isotope dilution mass spectrometry).
[c] Atmospheric contamination value quoted is the higher of duplicated ^{40}Ar analytical values.
[d] Calculated with constants of Steiger & Jäger (1977).
[e] Age from Hogarth & Roddick (1989).

Appendix III: Assignments of officers and men during and after the Frobisher voyages

In this listing we have used the following identifications: William Bendes (Cuba) = William Bennes (Baffin), Richard Gibs (Russia) = Master Gibbes (Baffin), John Hilliard (Brazil) = John Hellard (Baffin), Tege Hues (Brazil) = Tege Hewes (Baffin), Abraham Kendall (Roanoke) = Captain Kendall (Baffin), Henry Moyle (Cadiz and Ireland) = Captain Moyles (Baffin). Sources include Andrews (1959, 1972), Corbett (1898a, b), Gad (1971), Hakluyt (1927), Laughton (1895), Oppenheim (1902), Quinn (1955), Taylor (1959), Warner (1899) and, of course, *DNB* and the account books of Frobisher's northwest venture (*HL*, HM 715; *PRO* E 164/35, 36).

Name assignments	Frobisher voyages	Post-Frobisher naval or maritime
	Voyage, ship (duty or classification)	Year, sponsor or voyage (destination), ship (duty)
ALDAYE, James	1, unknown (sailor)	1579, King of Denmark (Greenland), unknown (admiral)
BEARE, James	2, *Michael* (master); 3, *Ann Francis* (pilot)	1584, Levant Company (Tripoli?), *Judith* (master) 1596, Lord Admiral (Cadiz), *Dreadnought* (master)
BENDES, William	3, *Michael* (mate)	1591, John Watts (Cuba), *Little John* (master)
CHANCELLOR, Nicholas	1, 2, 3, *Gabriel, Ayde, Judith* (purser, all 3)	1582, Fenton (Sierra Leone), Edward *Bonaventure* (purser)
COURTENAY, Thomas	3, *Armonell* (captain and owner)	1579, Lord Justice of Ireland (Smerwick), *Armonell* (captain)
DIAR[DYER], Andrew	2, *Ayde* (pilot); 3, *Hopewell* (pilot)	1584, Levant Company (Tripoli), *Jesus* (master)
DRAPER, Esdras	2, *Ayde* (sailor); 3, *Ayde* (sailor)	1582-3, Fenton (Sierra Leone & Brazil)*Galleon Leicester* (Steward)
ELLIS, John	3, *Ayde* (sailor)	1589, Chidley (Trinidad), *Wildman* (master)
EVANS, David	3, *Gabriel* (soldier & baker)	1582, Fenton (Sierra Leone), *Galleon Leicester* (baker)

Name	Frobisher voyages	Post-Frobisher naval or maritime assignments
	Voyage, ship (duty or classification)	Year, sponsor or voyage (destination), ship (duty)
FAIRWEATHER, Richard	3, *Beare Leicester* (master)	1582, Fenton (Sierra Leone & Brazil), *Bark Francis* (master)
FENTON, Edward	2, *Gabriel* (captain); 3, *Judith* (Lieut. Gen.)	1582-3, Fenton (Sierra Leone & Brazil), *Galleon Leicester* (admiral) 1588, against Armada (Channel), *Mary Rose* (captain)
FISHBOURNE, Richard	2, *Ayde* (boy); 3, *Judith* (sailor)	1587, Drake (Cadiz), *Little John* (captain)
GIBS, Richard	3, *Thomas Allen* (master)	1582, Muscovy Company (St. Nicholas), unknown (master)
HALL, Christopher	1, *Gabriel* (master); 2, *Ayde* (master); 3, *Ayde* (pilot)	1581, Muscovy Company (St. Nicholas), *Thomasin* (master) 1582-3, Fenton (Sierra Leone & Brazil), *Galleon Leicester* (master)
HAWL[HALL], John	2, *Ayde* (boy); 3 *Ayde* (boy)	1582-3, Fenton (Sierra Leone & Brazil), *Galleon Leicester* (sailor)
HILLIARD, John	3, *Ayde* (gentleman)	1582-3, Fenton (Sierra Leone & Brazil), *Edward Bonaventure* (sailor)
HUES, Tege	3, *Gabriel* (sailor)	1582-3, Fenton (Sierra Leone & Brazil), *Galleon Leicester* (sailor)
JACKMAN, Charles	2, *Ayde* (mate); 3, *Judith* (master)	1580-1, Muscovy Company (Kara Sea), *William* (vice-admiral)
JACKSON, Christopher	2, *Ayde* (trumpeter); 3, *Ayde* (trumpeter)	1582-3, Fenton (Sierra Leone & Brazil), *Galleon Leicester* (trumpeter)
KENDALL, Abraham	3, *Bark Dennis* (captain)	1585-6, Drake (Roanoke), unknown (master) 1589, Chidley (Trinidad), *Wildman's Club* (master) 1594-5, Dudley (West Indies & Orinoco), *Bear* (master & pilot) 1595-6, Drake & Hawkins, Panama, (*Defiance?*) (Master & pilot)

Name	Frobisher voyages Voyage, ship (duty or classification)	Post-Frobisher naval or maritime assignments Year, sponsor or voyage (destination), ship (duty)
MOYLE, Henry	3, *Francis* of Foy (captain)	1596, Essex & Howard (Cadiz), *Moon* (captain) 1599, Howard (Ireland), *Spy* (vice-admiral)
PEMBERTON, Robert	3, *Emanuel* of Bridgwater (miner)	1582-3, Fenton (Sierra Leone & Brazil), *Edward Bonaventure* (drummer)
ROBINSON, Edward	2, *Ayde* (sailor); 3, *Ayde* (sailor)	1582-3, Fenton (Sierra Leone & Brazil), *Galleon Leicester* (quartermaster)
ROBINSON, Peter	3, *Judith* (sailor)	1582-3, Fenton (Sierra Leone & Brazil), *Galleon Leicester* (sailor)
SALT, Richard	3, *Ayde*(?) (miner)	1582, Fenton (Sierra Leone), *Galleon Leicester* (sailor)
SMYTHE, John	3, *Emanuel* of Bridgwater (master)	1582-3, Fenton (Sierra Leone & Brazil), *Edward Bonaventure* (mate) 1588, against Armada (Channel), *Bark of Bridgwater* (captain)
WARD[WARDE], Luke	3, *Ayde* (gentleman)	1578, against pirates (Channel), unknown (captain) 1582-3, Fenton (Sierra Leone & Brazil), *Edward Bonaventure* (captain) 1588, against Armada (Channel), *Tramontana* (captain) 1588-9, patrol (Channel), *Tramontana* (admiral) 1591, patrol (Channel), *Swallow* (admiral)
YORKE, Gilbert	2, *Michael* (captain); 3, *Ayde* (captain)	1579-80, Parrot & Winter (Ireland), *Achates & Swiftsure* (captain) 1584, patrol (Channel), *Scout* (captain) 1595-6, Drake & Hawkins (Panama), *Hope* (captain)

INDEX

Page numbers in italic indicate illustrations or tables; numbers in parentheses indicate numbers of notes at the end of chapters.